Mathematical Structures of Nonlinear Science

Nonlinear Topics in the Mathematical Sciences

*An International Book Series dealing with Past, Current and Future Advances
and Developments in the Mathematics of Nonlinear Science*

Editor:

MELVYN S. BERGER

University of Massachusetts at Amherst, U.S.A.

Mathematical Structures of Nonlinear Science

An Introduction

by

Melvyn S. Berger

*Department of Mathematics,
University of Massachusetts at Amherst, U.S.A.*

KLUWER ACADEMIC PUBLISHERS

DORDRECHT / BOSTON / LONDON

Library of Congress Cataloging in Publication Data

Berger, Melvyn S. (Melvyn Stuart), 1939-
    Mathematical structures of nonlinear science / Melvyn S. Berger.
        p.   cm.
    Includes bibliographical references and index.
    ISBN 0-7923-0728-3 (alk. paper)
    1. Nonlinear theories.   I. Title.
QA427.B45   1990
003'.75--dc20
                                                            90-4817

ISBN 0-7923-0728-3

Published by Kluwer Academic Publishers,
P.O. Box 17, 3300 AA Dordrecht, The Netherlands.

Kluwer Academic Publishers incorporates
the publishing programmes of
D. Reidel, Martinus Nijhoff, Dr W. Junk and MTP Press.

Sold and distributed in the U.S.A. and Canada
by Kluwer Academic Publishers,
101 Philip Drive, Norwell, MA 02061, U.S.A.

In all other countries, sold and distributed
by Kluwer Academic Publishers Group,
P.O. Box 322, 3300 AH Dordrecht, The Netherlands.

Printed on acid-free paper

Printed in the Netherlands

Dedicated to my mother, Mrs. Hilda Heller Berger and
my aunt, Mrs. Anne Heller Jacobson

The research for this book was partially supported by a grant from
the AFOSR.

Table of Contents

Preface   ix

Chapter 1

Integrable Nonlinear Systems and their Perturbation   1

1.1   The Simplest Nonlinear Systems   2

1.2   Integration by Quadrature and Its Alternatives   17

1.3   Classical Mechanical Integrable Systems   38

1.4   New Ideas on Complete Integrability for Equilibrium
      Processes   40

1.5   Canonical Changes of Coordinates for the Mapping A   62

1.6   Bifurcation and the Integration of Nonlinear Ordinary and
      Partial Differential Equations   66

1.7   Qualitative Properties of Integrable Systems – Periodic and
      Quasiperiodic Motions of Dynamical Systems   74

1.8   Almost Periodic Motions of Dynamical Systems   113

Appendix 1   Nonlinear Fredholm Operators   118

Appendix 2   Bifurcation from Equilibria for Certain Infinite-
             Dimensional Dynamical Systems   122

Appendix 3   Elementary Facts about the Linear Dirichlet Problem
                                                                          127

Appendix 4   On Besicovitch Almost Periodic Functions   137

Chapter 2

General Principles for Nonlinear Systems   144

2.1   Differentiable Nonlinear Operators   145

2.2   Iteration of Nonlinear Operators   160

2.3   Nonlinear Fredholm Alternatives   172

2.4   The Idea of Nonlinear Desingularization   176

2.5   Variational Principles – New Ideas in the Calculus of
      Variations in the Large   192

2.6   Bifurcation   203

2.7   Bifurcation Into Folds   213

Table of Contents

Chapter 3

Some Connections between Global Differential Geometry and Nonlinear Analysis   **228**

3.1   Geodesics   231

3.2   Gauss Curvature and Its Extensions   242

3.3   Manifolds of Constant Gauss Positive Curvature   252

3.4   Mean Curvature   257

3.5   Simple Riemannian Metrics   261

3.6   Einstein Metrics   274

Chapter 4

Vortices in Ideal Fluids   **294**

4.1   The Early History of Vortices in Fluids   294

4.2   Formulation of the Vortex Concept in Ideal Incompressible Fluids   297

4.3   Axisymmetric Vortex Motions with and without Swirl   304

4.4   Variational Principles for the Stream Function for Vortex Rings without Swirl   312

4.5   Leapfrogging of Vortices   317

4.6   Vortex Breakdown   326

4.7   Nonlinear Desingularization and Vortex Filaments   342

Chapter 5

Mathematical Aspects of Superconductivity   **348**

5.1   The Simplest Nonlinear Yang-Mills Theory that Works   348

5.2   Physical Viewpoint   356

5.3   The Linear Approach to Superconductivity and Nonlinear Desingularization   359

5.4   Function Spaces for Symmetric Vortices   371

5.5   The Existence of Critical Points for I_λ Associated with Symmetric Vortices   376

Bibliography   **382**

Problems   **385**

Index   **415**

Preface

This is the first volume of a series of books that will describe current advances and past accomplishments of mathematical aspects of nonlinear science taken in the broadest contexts. This subject has been studied for hundreds of years, yet it is the topic in which a number of outstanding discoveries have been made in the past two decades. Clearly, this trend will continue. In fact, we believe some of the great scientific problems in this area will be clarified and perhaps resolved. One of the reasons for this development is the emerging new mathematical ideas of nonlinear science. It is clear that by looking at the mathematical structures themselves that underlie experiment and observation that new vistas of conceptual thinking lie at the foundation of the unexplored area in this field. To speak of specific examples, one notes that the whole area of bifurcation was rarely talked about in the early parts of this century, even though it was discussed mathematically by Poincaré at the end of the nineteenth century. In another direction, turbulence has been a key observation in fluid dynamics, yet it was only recently, in the past decade, that simple computer studies brought to light simple dynamical models in which chaotic dynamics, hopefully closely related to turbulence, can be observed.

The occasion to prepare this book was an invitation I received in 1987 to visit the Mathematics Department of the University of Southern California to deliver four lectures on the theme "Mathematics of Nonlinear Science" as part of their Distinguished Lecture Series on Mathematics. In this lecture series, given in January, 1988, I was free to choose any topics at all for each individual lecture, and I tried to give each lecture the individual flavor of my own ideas and research over the past twenty years, while at the same time, describing the contributions of other mathematicians in each area covered. I tried to pick four areas in which mathematics and science walk hand in hand, with each providing insights for the other. I view it as quite an astonishing fact that although the mathematical structures within mathematics have been known for thousands of years, it is only within the last few decades that patterns and deep analogies have been seen in fields that once seemed so different. For example, the calculus of variations and the vortex motion of fluids, or the obstructions to integrability caused by bifurcation processes.

I chose the four lecture topics in such a way that they would be interconnected and at the same time be capable of being understood independent of the other topics discussed. I also had the following goals in mind: simplicity of statement, historic links with mathematical and scientific culture, importance in future development of the subject, opportunity for new fundamental advances.

The four topics are as follows:

Topic 1 — New Ideas on Integrability of Nonlinear Systems

In this lecture I tried to analyze those nonlinear systems that have some kind of canonical form, to analyze why this canonical form cannot be perturbed, and how possible remedies for this fundamental problem can be approached.

Topic 2 — Vortex Phenomena in Fluids

Vortices have been observed in fluids for centuries, dating back to Leonardo da Vinci. Vortex phenomena have proved very basic in understanding fluid motions within the entire world of fluid dynamics, oceanography, and aeronautics, topics of nonlinear science par excellence. These seemingly disparate areas can be understood from a coherent viewpoint once mathematics is brought into the picture. For centuries, vortices in fluids could not be adequately treated, perhaps due to the lack of understanding of nonlinear science and its associated mathematical structures. Thus, the understanding of new mathematical ideas of nonlinear science proved a turning point in scientific discussion.

Topic 3 – New Connections between Global Differential Geometry and Nonlinear Analysis

For a number of centuries it was observed that geometric structures could be understood, and new insights obtained, by using the extremal characterizations that they often possess. Yet deeper insight into these geometric structures required advances in mathematical analysis that have occurred within the past twenty years. These problems often reduced to global questions in nonlinear partial differential equations. In addition, Einstein's theory of relativity provided the occasion for a fundamental change in the outlook for large-scale natural structures. All these questions involved nonlinear phenomena with an intricate, but comprehensible structure. I deal with some of these issues in this book.

Topic 4 – Superconductivity. The Simplest Nonlinear Gauge Theory That Works

The superconductivity of certain materials has proved an exciting development within the past few years. All over the world, large research teams are trying to discover new superconducting materials that may lead to the improvement of human life in manifold aspects. Prospects are very bright for this subject's potential advances. What are the mathematical structures that lie behind this subject? This is the question that formed the topic of the final lecture in my series. Amazingly, a number of issues discussed in the previous topics entered this field of study as well. Typical examples include vortex phenomena, nonlinear desingularization, calculus of variations, linearization processes, nonlinear eigenvalue problems, etc. . . .etc. I thought that this area was a fitting climax to the four talks described herein, exemplifying tremendous potential for the advancement of science and technology by nonlinear mathematical processes, building on inspired experimental investigation.

The present book ends with a collection of diverse problems, chosen to illustrate the material of the text or to fill in elementary facts assumed known to the reader. The subjects discussed in this book open a vast research panorama of mathematical research problems of sufficient depth, interest and difficulty to fill many pages; perhaps, in fact, another volume. I have included a few of these at the end of the problem collection. I owe many thanks to former doctoral students and colleagues for help with the text, in particular Y.Y. Chen, J. Nee, Professors Martin Schechter, Phil Church and Edward Fraenkel. Most of all, I owe more than I can ever convey in words, to my dear wife Diane, who among many other things, typed the entire text.

Northampton, Massachusetts
January 16, 1990.

Chapter 1 – Integrable Nonlinear Systems and their Perturbation

What are the simplest nonlinear structures that exist in science and mathematics? This question seems perhaps too grandiose to describe in any coherent way. However, this is precisely what is meant by an integrable nonlinear system. In terms of differential equations this topic classically involves questions of integrability by quadrature for ordinary differential equations, and for dynamical systems of mechanics, the notion of integrable systems goes back to the nineteenth century, when famous studies of Jacobi, Liouville, and Poincaré achieved great insight. The recent decade is quite unique in that two opposite directions of research come to the fore. The first direction involves extending systems to infinite dimensions, (i.e. systems with an infinite number of degrees of freedom). In the second extension, one analyzes simple dynamical systems depending on a real parameter, that are the negation of integrable. Such systems are called chaotic because, for example, the observed motions of such systems cannot be predicted with any accuracy over long time intervals. In this chapter I first ask the question, What nonlinear phenomena lie between these two extremes. Secondly, Can one create a mathematical language and structure to understand these diverse forms of integrability?

Section 1.1 The Simplest Nonlinear Systems

The study of nonlinear systems dates back to the beginnings of science thousands of years ago. Yet it was a special occasion when nonlinear structures became the focus of attention, as opposed to linear ones. Common sense is linked to the linear structures of science. Ratio and proportion, principle of superposition, effect proportional to cause, are all guideposts for linear thinking. Rule and compass constructions are generally linear in nature. What distinguishes nonlinear behavior in science? This topic is the focus of the present book. Among these nonlinear systems, it has always been an idea of key importance to distinguish the simplest ones.

In terms of algebraic equations, the simplest systems are classified by their degree. The major effects of the simplest nonlinear structure in this case are nonuniqueness of solutions of quadratic equations and the necessity to pass from the real number system to complex numbers, quite an achievement. When one focuses on differential equations, the situation is quite different. The simplest nonlinear differential equations have always been known as integrable. What does this term mean? The traditional meaning of an integrable differential equation is one that can be linearized after a restricted class of changes of variable. Thus the famous Bernoulli's equation is integrable but the Riccati equation is not.

§1. Complete Integrability in Classical Mechanics

For classical Hamiltonian systems of dimension 2N

$$(1) \qquad \dot{\beta}_i = \frac{\partial H}{\partial q_i}, \qquad \dot{q}_i = -\frac{\partial H}{\partial \beta_i} \qquad (i = 1, ..., N)$$

complete integrability implies integrability of the system (1) by quadrature after appropriate coordinate transformations. The appropriate coordinate transformations here

$$\begin{cases} P_i = P_i(p_i, q_i) \\ Q_i = Q_i(p_i, q_i) \end{cases}$$

are called "canonical transformations" and have as invariants the Poisson bracket

$$[f, g] = \sum \left(\frac{\partial f}{\partial p_i} \frac{\partial g}{\partial q_i} - \frac{\partial f}{\partial q_i} \frac{\partial g}{\partial p_i} \right)$$

of any two functions.

The criteria for complete integrability in classical mechanics is generally formulated in terms of separable systems relative to the Hamilton-Jacobi equations or Liouville's theorem. Both situations require the existence of N independent first integrals of the motions defined by (1) $f_1, \ldots, f_N$ whose Poisson brackets vanish. Such systems are said to be in "involution".

Assuming the set $M = \{(p, q) \mid f_i = c_i, i = 1, \ldots, N\}$ is a compact and connected smooth manifold, angle action variables (F_i, φ_i) can be introduced by global changes of coordinates, so that the system (1) can be reduced to

(2) $$\frac{dF_i}{dt} = 0, \qquad \frac{d\varphi_i}{dt} = c_i(F)$$

on M. This system is basically a linear one, since they imply F_i = constant and the angle action variables can be determined by an integration by quadrature. This approach can be extended to infinite dimensional Hamiltonian systems and partial differential equations. The celebrated Korteweg de Vries equation

$$u_t + uu_x + u_{xxx} = 0$$

is completely integrable in this extended sense, provided one fills in the analytic details with the inverse scattering method. A brief discussion of this topic follows. A more complete discussion can be found in the recent book of Drazin and Johnson. (See Bibliography). One key idea here is that the integrable systems discussed by this method all have a Hamiltonian structure.

   A more precise form of Liouville's theorem can be stated as follows:

Liouville's Theorem   Suppose the Hamiltonian system of dimension 2n described above has   n   independent conserved quantities $f_1, f_2, \ldots f_n$ in involution.   Moreover, assume the equations $f_i = c_i$ (c_i a constant,   i = 1, 2, ... n)   define a compact connected manifold M^c.   Then M^c is diffeomorphic to an n-dimensional torus, and the neighborhood of M^c is diffeomorphic to a direct product of this torus and $\mathcal{R}^n$.   Moreover, after appropriate coordinate changes (p, q) $\to$ (I, ϕ) the Hamiltonian equations can be written in the simplified form as

$$\dot{I} = 0$$

$$\dot{\phi} = \omega(I) \quad \text{where } \omega(I) = \frac{\partial H}{\partial I}$$

and the motion on M^c is quasiperiodic.

For a proof of this result, we refer to the book of V.I. Arnold, Mathematical Methods of Classical Mechanics. The study of almost all integrable Hamiltonian systems of mechanics are based on this result.

The N-dimensional linear harmonic oscillator given by the system

$$(*) \qquad\qquad \ddot{q} + Aq = 0$$

is perhaps the simplest integrable system. This mechanical system is said to have 2N degrees of freedom, since it represents a system of N second order equations. Here A is a real self-adjoint positive definite $N \times N$ matrix with positive eigenvalues $\lambda_1^2, \lambda_2^2, \ldots \lambda_N^2$, and $q(t)$ denotes a vector in $\mathbb{R}^N$. To find N conserved quantities in involution we let C be the constant orthogonal matrix diagonalizing A to the diagonal matrix D. Then setting $q = Cx$ we find, after a coordinate transformation

$$(**) \qquad\qquad \ddot{x} + C^{-1}ACx = 0 \quad \text{ where } C^{-1}AC = D$$

Let $x = (x_1, x_2, \ldots x_N)$, so that via $(**)$ the N desired conserved quantities can be denoted $\dot{x}_i^2 + \lambda_i^2 x_i^2 = \text{const.}$ $i = 1, 2, \ldots N$. It is easy to check that these conserved quantities are in involution.

By the definition given above, these quantities are independent and in involution. Hence Liouville's theorem can be applied to $(*)$ yielding an integrable system. The linear system $(*)$ is important for various reasons. First, it shows that a "coupled" dynamical system of ordinary differential equations can be simply uncoupled by a coordinate transformation the variables $\{q_i\}$ are connected by $(*)$

but unconnected in (**). This idea will be moved to the nonlinear case below. Secondly, it vividly shows how fragile the idea of "integrable" is to perturbation, since if a small higher order term in |q| is added to (*), in general, no simple uncoupling is possible. (However, see the Toda lattice example below for an exception). Indeed the perturbed system is nonintegrable ,no matter how small the non-zero perturbation.

Other standard examples include

I.   The usual heavy symmetric top fixed at a point on its axis. Such a system can be considered on $\mathfrak{R}^6$ and thus clearly has three conserved quantities: the Hamiltonian function H itself, and two projections of the angular momentum vector, first about the vertical axis, and then about the axis of rotation.

II.   Geodesics on an ellipsoid in three dimensional Euclidean space. Here one introduces with Jacobi special elliptic system of coordinates.   Then the so-called Hamilton-Jacobi method of integration operates, yielding three conserved quantities in involution.

III.   The N-dimensional Toda lattice whose Hamiltonian can be written as the sum of kinetic plus potential energies. For example, in three dimensions, we can write

$$H = \frac{1}{2}(p_1^2 + p_2^2 + p_3^2) + e^{q_1-q_2} + e^{q_2-q_3} + e^{q_3-q_1})$$

Here the Hamiltonian is of course preserved and, in addition, another constant, the momentum of the center of mass, here called P, depending only on the vector p variables.

It turns out, for example, in three dimensions, that there is a further conserved quantity K that can be written as a complicated function in terms of the (p, q) variables.

These conserved quantities are in involution and are independent. Thus their level sets form a compact connected manifold. Thus Liouville's theorem applies and the motion of the associated system is always quasiperiodic. This topic will be studied later in the Chapter.

As mentioned above, it is important to extend the notion of integrability to partial differential equations. Modern research has scored a real triumph here with the inverse scattering method applied to a number of time-dependent problems in one space dimension -- such as the Korteweg de Vries equation (denoted hereafter as the KdV equation)

$$(1) \qquad u_t - 6uu_x + u_{xxx} = 0$$

This equation has a number of remarkable properties that lead one to suspect the equation may have the structure of an "infinite dimensional" integrable Hamiltonian system for which an "infinite dimensional" analogue of Liouville's theorem may apply, even though such a result has never been established. The properties include:

(i) The KdV equation has an infinite number of conservation laws of the form $\qquad H_0(u) = \int_{-\infty}^{+\infty} u \, dx \ , \qquad H_1(u) = \int_{-\infty}^{+\infty} u^2 dx \ ,$

$H_2(u) = \int_{-\infty}^{+\infty} (u^3 + \frac{1}{2} u_x^2) dx \ ,$ etc.

(ii) A Hamiltonian formulation as $\dot{u} = D\left[\dfrac{\delta H_2}{\delta u}\right]$ where $\dfrac{\delta}{\delta u}$ is the Frechet derivative of H_2, i.e.

$$\frac{\delta F}{\delta u} = F_u - (F_{u_x})_x + (F_{u_{xx}})_{xx} \ \ldots$$

(iii) An analogue of a Poisson bracket

$$[A, B] = \int A_u D B_u dx$$

that satisfies Jacobi's identity. Here A_u and B_u denote the Frechet derivatives of A and B , i.e. $\dfrac{\delta}{\delta u} A$ and $\dfrac{\delta}{\delta u} B$ as above.

(iv) The infinite sequence of conservation laws $H_0, H_1, H_2, \ldots$ are in involution with respect to the bracket so that $[H_i, H_k] = 0$.

Among the key differences with finite dimensional Hamiltonian systems are the infinite dimensional phase space, the momentum variable p does not appear at all and the relevant skew symmetric matrix is replaced by the skew symmetric operator $D = \dfrac{\partial}{\partial x}$ with respect to the L_2 inner product. In fact at present no general infinite dimensional extension of Lioville's theorem is known.

In this case, complete integrability for the KdV equation can be proven by the "inverse scattering technique," a highly ingenious global linearization approach. This method is illustrated by the following diagram:

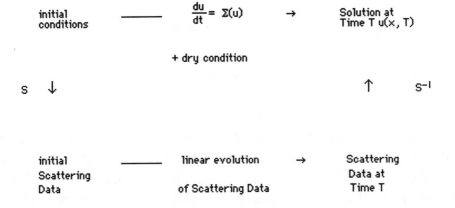

initial
conditions ———— $\frac{du}{dt} = \Sigma(u)$ → Solution at
Time T u(x, T)

+ dry condition

S ↓ ↑ S⁻¹

initial ———— linear evolution → Scattering
Scattering Data at
Data of Scattering Data Time T

FIGURE **MAPPING DIAGRAM FOR THE INVERSE SCATTERING METHOD**

Here S denotes the Forward Scattering Transformation and S⁻¹ denotes the Inverse Scattering Transform. The above picture denotes the schematic view of integrability of the initial value problem by the inverse scattering method.

Here an amazing linearization occurs since the scattering data of the linear evolution is associated with the spectra of the differential equation

(4) $Lu + \lambda u = 0$ + null dry conditions at infinity

Here Lu is the differential operator $Lu = u'' - f(x)u$.

The discrete part of this spectrum of L give rise to solitons and multisolitons. This discovery is remarkable and has caused tremendous excitement in present day mathematical research. Research credit for such discoveries belongs to many people, Kruskal and Zabusky, McKean, Novikov, Lax, among others.

It is rather remarkable that the number of equations of mathematical physics that can be studied by this inverse scattering method is rather large. In fact, there are at least seventeen examples of such equations. However, they all involve spatial dependence in one variable (with the notable exception of the KP equation). All these equations cannot be perturbed since the conservation laws necessary for their integration are destroyed. Moreover, it is extremely curious that the nonlinear equations involved can be completely studied by a linear problem, namely, the spectral properties of the operator L.

To solve the initial value problem for the KdV equation with $u(x, 0) = f(x)$ assuming $\int_{-\infty}^{+\infty} (1+|x|)\,|f(x)| < \infty$ one solves the linear scattering problem

$$\text{(S)} \qquad \psi_{xx} + (\lambda - f(x))\psi = 0 \qquad -\infty < x < \infty$$

where λ is an eigenvalue and determines the scattering data for (S) relative to the behavior of the eigenfunctions of (*), the discrete spectrum $-K_n^2$, the normalization coefficients $c_n(0)$ and reflection coefficients $b(k, 0)$. Remarkably, the time evolution of this scattering data is linear, the constants k_n remaining constant while the $c_n(t)$ and $b(k, t)$ satisfy simple first order homogeneous ordinary differential equations of the first order. The final step, in the inverse scattering method, S^{-1}, consists in the construction of a function $F(x, t)$ from the time evolved scattering data and using it to solve a certain linear integral equation, the Marchenko equation. The solution of the KdV equation at time t $u(x, t)$ by differentiation. The same inverse scattering procedure has been used to show a variety of time-dependent partial differential equations in one space dimension, for example

(i) the sine-Gordon equation $u_{tt} - u_{xx} + \sin u = 0$
(ii) the nonlinear Schrödinger equation $iu_t + u_{xx} + u|u|^2 = 0$.

Integration by Quadrature

The fundamental idea in any notion of integrable system is the notion of change of variables. An unravelling process, expressed mathematically, is known as "changing variables." In classical mechanics of the nineteenth century, this process was clearly defined by the words canonical coordinates and canonical change of variables. In the nineteenth century, one followed a definite procedure by introducing angular variables and conjugate variables in such a way that one reduced a Hamiltonian system to a simple equivalent linear model.

The crucial idea here is the notion of conserved quantity; namely, for physical systems with N degrees of freedom, an integrable system meant finding N conserved physical quantities, such as energy, angular momentum, etc. One thus looked at the N conserved quantities as an N – dimensional manifold , and then asked, when restrained to this manifold can one find such linearizing changes of coordinates? This is the content of Liouville's theorem mentioned above.

The direct way to proceed to a systematic understanding of these ideas is to begin with the simplest case of an N by N matrix A with real entries regarded as a linear mapping acting between two finite dimensional spaces of the same dimension N. The goal of this work with matrices is to solve as explicitly as possible linear equations of the form

(1) $Ax = y$

where y is a given N vector.

This is what integrability means for matrix equations of the type above. If these equations are written out, they involve a coupling of the variables involved in the components of x. The simplest integrable equations would be a diagonal one where the equations are all uncoupled so that one could solve one equation after another and thus find the solutions to the matrix equation (1) explicitly. Thus, instead of solving N coupled equations in N unknowns, one would solve N uncoupled equations in the same N unknowns. The equations could be written

(2) $c_i x_i = y_i$ $(i = 1, 2, \ldots N)$

It turns out that every linear system of the form (1) is integrable in the sense that we can make two invertible linear changes of variables, one for the x variables and one for the y variable that uncouples the matrix equation (1) and to the system (2). This is a well-known result of linear algebra and is known as the rank normal form. This normal form for the matrix A has a special property of stability. This means that if the entries of matrix A are slightly perturbed, the rank normal form remains unchanged. The rank normal form is defined by saying that two square matrices A and B are equivalent if one can find two invertible changes of coordinates called P and Q such that $A = PBQ$. It is known from linear algebra that the only invariant for this notion of equivalence is the rank of the matrix. (It is this idea that can be extended to the nonlinear case, provided one makes a distinction between local and global changes of coordinates). This means that A and B have the same rank, that they are equivalent. Moreover, if rank of A is K, B can be chosen to be a diagonal matrix where diagonal entries consist of K ones and N – k zeroes. This notion of stability is very important in numerical analysis, because in that case computer computations require truncation, and (in order to compute accurately with matrices, a stable normal form is required. Thus the rank normal form, or a small variation of it, has great practical significance.

The usual method of analyzing this linear problem involves the well-known question of conjugacy of matrices. That is, can one find a simple change of coordinates for the x variable and the y variable that reduces the complicated system (1) to the uncoupled system (2)? One then reduces the mathematical question to one involving the diagonalization of the matrix A. Explicitly put, one asks, under what circumstances can one find a linear change of coordinates P such that PAP^{-1} is diagonal?

If this question can be answered in the affirmative, and it can be in a number of cases that one can mention, the normal form for A consists of a diagonal matrix where diagonal entries are the eigenvalues of the matrix A. Cases for which this question can be answered in the affirmative are all self-adjoint matrices A and all matrices that possess N distinct eigenvalues. The associated normal form is called the Jordan canonical form. For our purposes, it has one major drawback, namely, it is not stable under a small perturbation. Namely, unlike the rank normal form, it is not stable under a small perturbation of its entries. However, it is just this notion of conjugacy that is relevant for the inverse scattering method discussed above.

However, there is even a simpler basic example, that of integration by quadrature. In fact, integrable dynamical systems as described above can be viewed as extensions of this notion of integrability. This notion goes back to Newton in the seventeenth century. In the very beginning of calculus, are studies of nonlinear differential equations of the forms

(3)
$$\frac{dy}{dx} = f(x, y)$$

$$y' = f(x, y) \qquad \text{where } y' = \frac{dx}{dy}$$

One asserts that this equation is integrable by quadrature if one can find changes of variables $x \rightarrow X(x,y) \quad y \rightarrow Y(x,y)$ that reduce the system to the form

(4)
$$\frac{dY}{dX} = g(X)$$

Note here, that more than one change of variables was allowed in these historical studies, and that a number of simple ordinary differential equations of the form (1) were proven not to be integrable by quadrature. The simplest among these was the equation of Riccati that can be written in the simple form

(5)
$$\frac{dy}{dx} + y^2 = f(x)$$

Liouville himself proved, in the nineteenth century, that for a large number of functions $f(x)$ that this equation could not be reduced to the linear form as required in the definition. One way to find systems integrable by quadrature in this classical sense is to consult well-written books on the calculus. Yet the question remains, why the simple Riccati equation (5) is not integrable by quadrature. This question and its extensions will be discussed in the subsequent sections of this chapter. In fact, when properly defined, one can show that the Riccati equation is integrable in a certain extended sense. This extended sense of integrability will now be explained utilizing a modern mathematical viewpoint.

Section 1. 2 Integration by Quadrature and Its Alternatives

It is amazing that one can use modern mathematics to go well beyond this classical notion of integrability. This fact is one of the themes of this Chapter. One considers an updated notion of integrability by quadrature by considering a nonlinear extension of the idea of rank normal form mentioned above. Briefly put, the idea is as follows:

One considers the right hand side of the differential equation (3) as an operator A acting between linear spaces of the same dimension called hereafter X and Y. One attempts to find smooth invertible changes of coordinates C_1, C_2 such that the composition $C_1 A C_2$ reduces to an uncoupled system, and moreover, that this reduction to an uncoupled system is stable under small perturbations in the form of the equation (3). Moreover, we would like to diagnose exactly how an idea goes beyond the classical historic one mentioned above.

It turns out that the Riccati equation is a test case for this new idea. In fact, the problem with the classical idea of integration by quadrature, from our viewpoint, lies in trying to reduce all differential equations of the form (3) to the form (4) in order to be integrable. The problem with this idea is that it avoids letting nonlinearity play any role in the notion of integrability by quadrature. The simplest nonlinear effect for the Riccati equation is bifurcation. This bifurcation shows that nonlinear effects must be taken into account for any comprehensive integrability idea to work. To exhibit bifurcation in the Riccati equation, we simply ask to find periodic solutions of any fixed period for the equation

(6)
$$y' + y^2 = c$$

where c is a real parameter varying between positive and negative numbers. We shall show, by a very simple argument, that for any fixed period P the Riccati equation has no nonconstant periodic solutions. Thus, as c varies from a negative constant to a positive constant, the real periodic solutions of fixed period T varies between 0, 1and 2, and that the only one with a unique solution occurs when $c = 0$ in which case $y(t) \equiv 0$. Here is a proof of these simple facts.

Assume that the equation (6) has a periodic solution y of any fixed period, P say. Multiply both sides of this equation by dy/dx and integrate over a period P, obtaining

(7)
$$\int_0^P y'^2 dx + \int_0^P y'y^2 dx = \int_0^P y'c\,dx$$

The last two integrands in this equation are both exact differentials, and so the corresponding integrals are identically 0 when periodic boundary conditions are taken into effect. This reduces (7) to the statement that the first integral on the left hand side of (7) vanishes. This clearly implies that the derivative of $y'(x) = 0$. Hence the equation (6) reduces to the statement that y = a constant, from which the stated result follows. Clearly this implies that bifurcation in the Riccati equation occurs at $y(x) = 0$, since the number of real periodic solutions with fixed period P changes exactly there as the parameter c changes from a negative to a positive value.

The Diagonalization of Nonlinear Differential Operators —
A New Approach to Complete Integrability

One goal in studying problems of nonlinear science consists of formulating a given problem in terms of nonlinear systems, generally, systems of nonlinear differential equations (ordinary or partial) supplemented by boundary conditions. One then attempts to integrate these equations as explicitly as possible using all the resources of mathematical analysis, geometry, algebra, and current large scale computing. In fact, as mentioned above, the goal of explicitly integrating nonlinear differential equations goes back to the origins of calculus with Newton, and has proved to be of crucial importance in subsequent centuries. In this section, we wish to exhibit some new ideas for integrating nonlinear differential equations supplemented by boundary conditions by using modern mathematical analysis and topology. We will focus on only the simplest problems that require this new conceptual framework, but with enough explicit detail so that the problems studied can be calculated explicitly on present-day computers. Hopefully, this will be only the first step in this new theory, and other researchers will add key new developments to enhance these ideas.

Our discussion is split into several parts. First we discuss the goals and conceptual details of a new theory of integrability, especially as it differs from the contemporary inverse scattering approach to complete integrability. Then we review some of the history of the subject of the explicit integration of nonlinear differential equations. In the third section we discuss details concerning the notion of diagonalizing a nonlinear operator, and compute some invariants for this problem in terms of explicit examples. These explicit examples will be chosen so that the notion of global linearization breaks down, but the nonlinear phenomena involved exhibit a simple intrinsic structure, both geometric and analytic. Then we show how this work leads to a systematic procedure for explicitly solving nonlinear differential equations of varying types.

A New Theory of Complete Integrability

There are a number of desirable properties of a new theory of integrability for nonlinear differential equations. They include

(1) **Independence of Space Dimension**   (The usual theories have been limited to partial differential equations of one-space dimension).

(2) **Stability**   Under a small perturbation a differential equation integrable in this sense should be stable in that the methods used for explicit computation are not totally destroyed, but on the other hand should lead once more to the explicit solution of the perturbed problem.

(3) **Inclusion of Classical Examples** (not integrable by quadrature). The theory we describe applies to both ordinary and partial differential equations. We wish to rethink the classical notion of integrability by quadrature in terms of this new point of view. In doing this, examples that were not integrable by quadrature should turn out to be integrable by these new methods.

(4) **Genuinely Nonlinear Phenomena** The current examples of complete integrability do not involve bifurcation, i.e. cases where prediction by linearization breaks down, and in fact, are based on "global linearizations." The inverse scattering method is a case in point.   The theory we shall describe will include bifurcation

phenomena in an instrinsic manner, and so will go beyond the linearization idea.

(5)   **A Systematic Procedure**   The theory we begin to describe can be checked for complete integrability by computing certain "invariants," the first step in a systematic procedure for integrability. These invariants are intrinsic, and are necessary conditions for this notion to work.   In addition, we shall use the ideas of functional analysis, taken in a nonlinear context, as a unifying feature for these problems.

(6)   **Connection with Existence Theory**   The methods we shall describe build on and sharpen the existence theory for nonlinear differential systems based on topological invariance such as the degree of a mapping, minimax determination of saddle points, calculus of variations, etc.   In so doing we show that the integrable systems that we describe cannot be treated by these existential arguments.

The basic idea of our notion of complete integrability begins with the following simple observations:

Suppose a differential equation of the form

(1) $Au = g$

is regarded abstractly as an operator equation between two real Banach spaces, X and Y; that is, the mapping A has domain X, and range Y. Suppose all the singular points S and singular values $A(S)$ can be computed explicitly for the mapping A, where the technical terms singular points and singular values will be defined below. We assume A to be a Fredholm operator of index 0 in the sequel so the singular points S = {x: A'(x) is not invertible} and the singular values A(S) is the image under A of the set S. The significance of these sets S and A(S) for this new theory of integrability is that bifurcation phenomena are crucial for this new integrability concept, and that the geometry of the sets S and A(S) are a measure of the complexity of the problem's bifurcation. Then what more need be said to integrate to solve equation (1) explicitly? Notice that the parameter dependence on A has not been explicitly defined.

The following picture is helpful. We summarize its content by saying that the mapping A is equivalent to C if there are global diffeomorphisms h_1 and h_2 as indicated below, such that

(2) $h_2A(x) = Ch_1(x)$ for all x in the Banach space X.

(Sometimes the notion of a diffeomorphism can be weakened to that of homeomorphism so that we do not require h_1 and h_2 to be differentiable).

$$
\begin{array}{ccc}
 & A & \\
X & \rightarrow & Y \\
 & & \\
h_1 \downarrow & & \downarrow h_2 \\
 & & \\
X_1 & \xrightarrow{} & Y_1 \\
 & C & \\
\end{array}
$$

Diagram   **MAPPING DIAGRAM FOR GLOBAL NORMAL FORMS**

This picture signifies that A is an operator between the spaces X and Y, h_1 and h_2 are global coordinate changes between the Banach spaces X and X_1, and Y and Y_1, and C is the canonical map between X and Y. The simplest canonical map would be a diagonal operator which can be described as follows. Suppose X_1 and Y_1 are denoted $(x_1, x_2, x_3, ...)$, $(y_1, y_2, y_3, ...)$, then the diagonal map D would be defined as $D(x_1, x_2, x_3, ...) = (f_1(x_1), f_2(x_2), f_3(x_3), ...)$.

Now the diagram clearly identifies what will be the invariants for diagonalizing a nonlinear operator exactly in the same manner as the rank of a matrix is the invariant in the linear case. The invariants for diagonalizing the operator A will be the singular points and singular values of the operator A as a mapping between the Banach spaces X and Y. Indeed, these points and their structures are preserved under the global changes of coordinates h_1 and h_2. In the case of diagonalizability, the canonical map C will be a diagonal map as defined above.

Invariant Properties for A

In order to discuss our ideas on explicit solutions with some degree of generality, it is of great interest to discuss the invariative properties of the operator A. These properties will be so chosen that they are independent of changes of coordinates in the Banach space context. These properties will have the additional virtue of "stability." This means that when the operator A changes to Ã, these properties do not change. This is one of the key ideas of the operator approach to nonlinear analysis. Here is an outline of some of the key invariant properties we can use.

1. **Proper mapping property**   A mapping is called proper if the inverse image of a compact set under the mapping A is also compact. This property will be of fundamental importance in determining what happens in passing from the local to the global aspects of the mapping A.

2.   **Singular points of A and their geometry** (Bifurcation points). A very important infinite dimensional manifold connected with A is a set of points at which the inverse function theorem breaks down. Such points are generally called bifurcation points of the mapping A. Under our equivalence diagram we note that these points denoted here by S are mapped homeomorphically by h_1 onto the singular points of the canonical map C. Since one can compute the set S for A and for C explicitly, we will be able by this means to construct at least partially the mapping h_1. This study differs from previous approaches in that we need to study the global geometry of the set S. For example, the natural questions arise: is S connected, is S a manifold, is S homeomorphic to a hyperplane of co-dimension one?

3. **Singular values of A, A(S), and their geometry** The set of images of S under the mapping A is called the set of singular values of A and denoted by A(S). For a broad class of mappings A it is known that the number of solutions of equation (1), Au = f, for f fixed, is constant in each component of Y − A(S) provided the mapping A is proper. This result gives us a clue about how to construct explicit solutions based on the mapping A; namely, we must compute the infinite-dimensional manifold A(S) and determine the basic geometry of this set. For this set determines precisely how the solutions of equation (1) change. Moreover, the topology of the set A(S) is invariant under the change of coordinate h_2, so that the singular points of the canonical map

C, C(S), and A(S) must be homeomorphic. This property in turn determines the global coordinate transformation h_2.

4. **Fredholm mapping property**   An operator A has the Fredholm mapping property if its Frechet derivative is a linear Fredholm operator between the spaces X and Y. (See Appendix I of this chapter). This means that the linear mapping has a finite dimensional kernel, a finite dimensional co-kernel, and a closed range. Again this property is invariant under the change of coordinates h_1 and h_2, so that if A is a Fredholm operator, so is C. Moreover, we generally require that A will be a Fredholm operator of index zero, i.e. we require that dim ker $A'(x)$ = dim coker $A'(x)$ for each x. This property is crucial at a fixed point x for the linear Fredholm theorems to be valid. What we shall achieve by using geometry is to generalize these results to a global context.

5. **Local classification of singular points**   H. Whitney and R. Thom studied the local classification of singular points under changes of coordinates in finite dimensions. In so doing, they obtained local normal forms for the mapping A near a singular point. Our goal is to extend this work in two ways: to an infinite-dimensional context of Banach spaces and secondly, to a global context. The possibility of doing this work is simplified because we shall study some classic operators A of mathematical analysis and its applications. Although these operators have been studied for some time, the methods we suggest are <u>intrinsic nonlinear approaches</u> instead of the usual linearizations.

For example, one asks the following question: (Q): Given a singular point x   of a nonlinear (index 0) Fredholm operator A acting between real Hilbert spaces   X   and   Y, when is   x   a generalized (Whitney) fold of infinite dimension?

To answer this question, we need to consider the second term in the Taylor series of   A   expanded about   x   and to determine its nondegeneracy. For example, assuming   A   maps between Hilbert spaces, we call a Whitney fold of a C^2 nonlinear Fredholm operator of index zero   A,   a singular point   x   of   A,   such that after a local coordinate change,   A   can be written (near x) as $(t, v) \to (t^2, v)$, where $t \in \mathfrak{R}$ is the coordinate in the direction   ker A'(x)   and   v   is the associated orthogonal complement.   For example, we proved (together with P. Church)

Theorem On Infinite Dimensional Whitney Folds   Let   A   be a C^2 Fredholm map of index zero between two Hilbert spaces H_1 and H_2 with a singular point at   x. Suppose

(i)   dim Ker A'(x) = 1   and
(ii) $(A''(x) (e_0, e_0), h^*) \neq 0$,   where $e_0 \in$ Ker A'(x)   and $h^* \in$ Ker [A'(x)]*   are both not identically zero.

Then   A   is a local Whitney fold near   x.

This result is easily extended to real Banach spaces   X , but for our purposes Hilbert space formulations suffice. Its proof, in fact a global version, will be given later in this Chapter.

It turns out that in the theory of differential equations we shall be able to answer this integrability question in a number of cases, and the ideas going into the theory are an extension of bifurcation theory and linearized stability theory of applied mathematics.   Our idea is that the mapping   A   is equivalent to a canonical mapping C called "the global normal form" for A after changes of coordinates. The question then becomes one of determining how to compute the global coordinate changes, and what are the appropriate normal forms for the operator A?

We ask the question, what <u>nonlinear</u> differential operators A can be globally diagonalized by similar notions?

Thus the procedures leading to diagonalizability of a nonlinear operator A can be listed as follows.

(1) Compute all singular points S for the mapping A and determine the geometric structure of S as a submanifold of X.

(2) Compute all singular values for the operator A(S) and determine the geometric structure of A(S) as a submanifold of Y.

(3) Compute the same singular points Σ and singular values D(Σ) for the desired diagonal map D and determine their geometric structures.

(4) Find global changes of coordinates that map S onto Σ and A(S) onto D(Σ).

Thus the question of diagonalizability becomes a combination of geometry and analysis. By this I mean ,analysis is needed in the selection of the Banach spaces X and Y, and the determination of the singular points S and A(S). These turn out to be interesting questions of boundary value problems for differential operators. The geometric part of the problem consists of determining the geometric structures of the singular sets in question and determining the appropriate changes of coordinates. In this book for most of the discussions, we restrict attention to the simplest genuine nonlinear case in which all singular points in question are Whitney folds. However, many of the arguments are valid more generally. In the sequel, we shall demonstrate that even by restricting attention to this simple case, many interesting nonlinear problems can be considered as integrable. These cases include the following:

1) The classic Riccati equation with periodic forcing term

2) Certain nonlinear elliptic Dirichlet problems defined on Euclidean spaces of arbitrary dimension

3) Nonlinear parabolic partial differential equations with an inhomogeneous forcing term that depends periodically on the time variable.

Perhaps this is a good place to compare our ideas with those of the inverse scattering method of complete integrability of differential equations. This theory has been immensely successful in discussing evolution equations in one-space and one-time dimension of the form

$$(3) \qquad\qquad u_t = Lu + Nu$$

However, this theory has had little success with problems involving higher dimensions partially because, I believe, the theory requires an infinite number of independent nontrivial conservation laws for its applicability. The theory is based on the equivalence of the system (3) to a global linearization of the form (4) below.

$$(4) \qquad\qquad u_t = Lu$$

Here L is a linear differential operator involving only the spatial derivatives, and N is the nonlinear part of the equation (3). This approach avoids bifurcation phenomenon in integrable nonlinear problems. This contrasts with our goals since bifurcation phenomenon are the intrinsic invariants that we utilize for diagonalizing a nonlinear differential operator. Indeed, the singular points that we utilize in our research are precisely the bifurcation points of the associated nonlinear differential operator. This theory of inverse scattering and the invariants associated with it, namely the conservation laws, are generally destroyed under a perturbation. Thus any theory of stability associated with complete integrability in this sense requires an entirely new analysis. As mentioned at the very beginning of this section, our approach includes a stability analysis in its intrinsic formulation. In addition, as we shall point out in later sections, our theory relates to nonlinear differential equations in all space dimensions.

In fact, the usual arguments of applied mathematics is relevant here. Due to experimental error, the nonlinear differential equations defining a physical problem are not known precisely. Thus since we suppose diagonalizability of A is an intrinsic property of the equation (1), and the physical problems associated with it, some such stability property is a necessary point of our new theory.

A Short History of Explicit Integration of Differential Equations

In the eighteenth century the notion of "integrability by quadrature" was the key approach to explicit integration. In this theory, one begins with a differential equation

$$(5) \qquad \frac{dy}{dt} = f(y, t)$$

and attempts, by changes of coordinates of the form $y = y(Y, T)$ and $t = t(Y, T)$, to reduce (5) to the inhomogeneous linear equation

(6) $$\frac{dY}{dT} = g(T)$$

This method proved immensely successful. However, not all ordinary differential equations could be analyzed by this means. A simple example, such as Riccati's equation

(7) $$\frac{dy}{dt} = y^2 + t^n$$

could be integrated by quadrature for a countable number of n's, but could not be integrated by quadrature for other values of n (see Watson, <u>Theory</u> <u>of</u> <u>Bessel</u> <u>Function</u>, for details). This is also mentioned in Arnold's recent book, mentioned in the Bibliography. This theory was amplified by Ritt forty years ago, and became the subject of differential algebra.

Another approach of the nineteenth century, already mentioned above, was due to Liouville. Again, as already mentioned, in this approach one considered canonical Hamiltonian systems of dimension 2N, and showed that if such systems had N independent conservation laws in involution, then the system could be integrated by quadrature. The method of inverse scattering can be viewed as an extension of this method to the case when $N \rightarrow \infty$. E. Cartan studied the problem of complete integrability locally by using differential geometry of connections and Lie theory (see Arnold's book for recent survey). However, this method was purely local and requires a great deal of additional research. In fact, earlier, Sophus Lie studied integrability of ordinary differential equations as an application of his ideas on symmetry. In all these cases coordinate transformations preserving the Hamiltonian structure are a well developed and useful aspect of nonlinear science. Our methods are

slightly different because they use the notion of functional analysis
and Fourier analysis in their formulation.

Determination of Singular Points and Singular Values and their Classification

A key problem in our approach to integrability is the
determination of singular points and singular values of a
differential operator.   Here we outline a general method for this
determination and illustrate it by examples.

Suppose $g(u)$ is a C' function of its arguments and set

$$(8) \qquad\qquad Au = Lu + g(u)$$

where L is a linear (partial) differential operator supplemented by
linear boundary conditions (assumed to be a linear Fredholm operator
of index 0 acting between Banach spaces X and Y). To find the
singular points of S of A we attempt to find the nontrivial
solutions (v) of the linearized equation $A'(u)v = 0$ at $u = \bar{u}$, i.e.,

$$(9) \qquad\qquad Lv + g'(\bar{u})v = 0$$

Letting $\bar{u}$ is such a value, then $\bar{u} \varepsilon S$ and the singular set S is the
set of such point $\bar{u}$ where (9) has a nontrivial solution. Moreover,
assuming the real valued function $g'(t)$ has a single-valued inverse
G, we have the following formula for $\bar{u}$ from (9)

$$(10) \qquad\qquad \bar{u} = G\left(-\frac{Lv}{v}\right)$$

Here we suppose $\dim \ker(L + g'(\bar{u})) = 1$. Then the singular values of
the differential operator A defined by (10) are given by the formula

(11) $A(\overline{u}) = L\overline{u} + g(\overline{u})$,

i.e.,

$$A(u) = LG(-\frac{Lv}{v}) + gG(-\frac{Lv}{v}) \quad \text{from (10)}.$$

Example 1. Riccati's Equation

As an example of this approach we consider Riccati's equation

(12) $u' + u^2 = g(t)$ with $u(0) = u(T)$

where $g(t)$ is a L_2 T-periodic function, for fixed T. We consider only T-periodic solutions and so regard

(12') $Au = u' + u^2$,

called the Riccati operator in the sequel, as a mapping between the Sobolev spaces of periodic functions $H_1(0, T)$ into $L_2(0, T)$.

First we observe that A (as defined above) is a nonlinear Fredholm operator of index 0. To this end, we compute $A'(x)y$ near an arbitrary point x, $A'(x)y = \frac{dy}{dt} + 2xy$ with the same T-periodic boundary conditions for y. Thus, via the Sobolev imbedding theorems, $A'(x)y$ can be represented as the sum of an invertible linear map plus a compact map. Thus $A'(x)$ is a Fredholm linear operator of index zero. Thus one easily shows A is a smooth nonlinear Fredholm operator of index 0 between these Hilbert spaces (cf. Appendices 1 and 3 at the end of this chapter).

One can easily prove the mapping A is proper, i.e. the inverse image of a compact set, by using an elementary version of the Sobolev-Kondrachev theorem. One simply squares both sides of (12') and integrates both sides over the period (0,T), using the periodic boundary conditions. One then obtains relative to the L_2 norm

$$\|Au\|^2 = \|u'\|^2 + \|u^2\|^2$$

Note here the cross product term vanishes by virtue of the periodic boundary conditions.This equality immediately yields the desired properness.

Moreover, an easy computation yields

(13) $A'(u)v = v' + 2uv$

Nontrivial T-periodic solutions for the equation $A'(u)v = 0$ occur (as is easily shown,see Section 2.7 for a more general case) exactly when the mean-value

(14) $$\int_0^T u(t)dt = 0$$

Thus, from the viewpoint of infinite dimensional geometry, the set of singular points for the operator Au is a hyperplane of codimension one on H_1 (the set of all Fourier series in the Sobolev space without constant term). By using our criterion for Whitney folds mentioned above, it is easy to see that each singular point $u(t)$ of the Sobolev space is a fold. To verify this note by Taylor series expansion,

$$A(x + h) = A(x) + [h' + 2xh] + h^2$$

This shows that the second derivative of A at x is "nondegenerate," so we will find that every singular point of A is a fold and so has the desired local normal form. To this end, we simply check that if u is a singular point of A,

(i) dim ker A'(u) = 1, since first order linear ordinary differential equations cannot have two linearly independent solutions.

(ii) $(A''(u) (e_0, e_0), h*) > 0$, where $e_0 \in$ Ker A'(u) and $h* \in$ Ker $[A'(u)]^*$. The reason for this is that the inner product occurs in the L_2 space, so that it is necessary only to compute the product of A''(u) (e_0, e_0) and h*. To this end we note that h* is not identically zero and so can be chosen to be strictly positive, and in the same way A''(u) (e_0, e_0) does not depend on u for the Riccati operator (as is shown above) and thus also can be chosen to be strictly positive.

Thus, more explicitly, to verify $x \in$ S(A) is a fold we note that

$$(A''(x) (e_0, e_0), h*) = \int_0^T e_0^2(t)h*(t)dt$$

and this integral is necessarily nonzero since h* is a nonzero solution of the equation $\frac{dy}{dt} - 2xy = 0$, which necessarily is of one sign.

To find the singular values we follow the prescription given above with $Lu = u'$ and $N(u) = u^2$ so that $Gu = \frac{1}{2}u$. Then, by (11), and setting $v = w^{-2}$ so $\bar{u} = \frac{w''}{w}$ we find

$$(15) \qquad A(\bar{u}) = (\tfrac{w'}{w})' + (\tfrac{w'}{w})^2 = \frac{w''}{w}$$

(after simplification). Thus $A(\bar{u})$ satisfies the Hill's equation

$$(16) \qquad w'' - A(\bar{u})w = 0$$

with $w > 0$, and so we can apply the elementary spectral theory for this situation so that w is the positive eigenfunction of $Lw = w'' - A(\bar{u})w$ with associated eigenvalues $\lambda_1(A(\bar{u})) = 0$. Thus, the singular values of A, $A(S)$ consist of all $g \in L_2(0, T)$ for which $\lambda_1(g) = 0$. Symbolically we suppose

$$(16') \qquad Lg(w) = w'' - g(t)w$$

has first eigenvalue $\lambda_1(g)$ and we write

$$(17) \qquad A(S) = \{g \in L_2, \; \lambda_1(g) = 0\}$$

It turns out, thus, that $A(S)$ can be studied as an infinite dimensional manifold in a Hilbert Space. In fact $A(S)$ is a smooth, convex surface of codimension one in L_2 dividing this Hilbert space into two connected pieces. Thus for this example we have reached the situation mentioned in our abstract discussion..

In order to proceed further, it is necessary to work with the singular points that arise in the above example. In fact, we shall find a local normal form for each singular point that we have computed, and moreover obtain a local diagonalization of the operators involved near each singular point. To this end, we shall extend the classification of singular points as folds due to Whitney. In particular, previously, we extended the notion of a Whitney fold to infinite dimensions.

In the above example, it is interesting to see in just what cases the singular points are folds. In fact it turns out that A is a global fold, and so A can be globally diagonalized.

We state this result formally in the following terms.

Theorem   (McKean–Scovel)   Let the Riccati operator $Au = u' + u^2$ (subject to T-periodic boundary conditions) acting between the Hilbert spaces $H_1(0, T)$ and $L_2(0, T)$. Then there are explicit smooth global coordinate changes h_1 and h_2 such that (relative to appropriate Fourier series decompositions of H_1 and L_2)

(19) $$h_2^{-1} A h_1(t, w) = (t^2, w)$$

(i.e. A is a global fold).

Here   explicit   formulae   for   the   global   coordinate transformations can be written, for example,

(20) $$h_1(t, w) = u(w) + t \exp(-2 \int_0^x u(s)ds)$$

Here u(w) denotes a function in S with coordinates (0, w) (notice that as required when t = 0 , h_1 maps the singular points of A into the singular point of the canonical global fold map. It is easily shown that h_1 is a global homeomorphism of H_1 into itself).

We shall not give the remaining details concerning the global changes of coordinates here, but defer to the paper of McKean-Scovel mentioned in the Bibliography. In the next few Sections, we shall again find a global Whitney fold for another example defined on a bounded domain of arbitrary finite dimension. In this case, however, we shall give all the details for the global coordinate transformations.

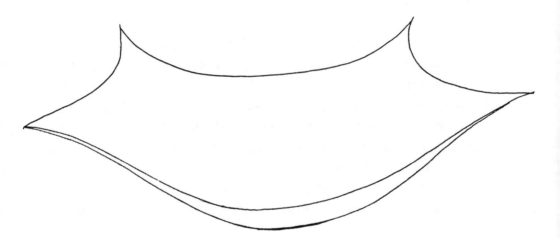

The manifold A(S) divides the Hilbert space into two connected pieces.

Section 1.3   Classical Mechanical Integrable Systems

In considering nonlinear partial differential equations with a Hamiltonian structure, it has been customary to consider nonlinear evolution equations of the form

(1) $$\frac{\partial u}{\partial t} = \Sigma(u)$$

supplemented by boundary conditions on the spatial components of u. Then one attempts to study the initial value problem for (1) by making a formal change of coordinates to simplify (1), hopefully to a problem with linear structure.

This was accomplished for the Burger's equation in the 1930's. See the following problems showing a global linearization for the associated initial value problem. This linearization idea has been extended greatly for nonlinear evolution involving one spatial dimension via the inverse scattering method. This method requires an infinite number of independent conservation laws and instead of focussing on the Poisson Bracket formulation one rewrites the nonlinear evolution equation in terms of the commutator formalism of P. Lax so that the nonlinear evolution equation has the form

$\frac{\partial u}{\partial t} = [L, B]$ where $[L, B]$ denotes the commutator of L and B.

Here L and B represent differential operators. Here the following diagram illustrates the inverse scattering method with scattering transformation S and appropriate S inverse scattering method.

This leads to the conclusion that there must be a whole new theory of integrability for higher dimensional nonlinear evolution equations since these equations of mathematical physics generally do not have an infinite number of conservation laws. The next step in my thoughts was the realization that bifurcation was excluded from the ideas of integrability in both the ideas of Section 2. In the remaining sections I shall outline my new integrability idea in fairly loose terms. The interested reader can find more information in some of the articles listed in the bibliography. Because the audience for this book is not primarily composed of mathematicians, I have relegated the few proofs contained here to a small appendix. Once again, many of the other proofs will be found in the articles of the bibliography. However, it should be added that the glory of mathematics is in its proofs and its ideas so that proofs clarifying the concepts involved are absolutely essential to keep a researcher from the pitfalls of error.

Two difficulties arise in this connection however. First, higher dimensional examples of nonlinear partial differential equations, integrable by this means have been very difficult to find and secondly, stability arguments associated with Liouville's motion of complete integrability require a radically new conceptual framework (e.g., KAM theory). Moreover, it is believed by many that the only integrable systems are those that can canonically be transformed to linear ones.

Furthermore, infinite dimensional systems (integrable in the Liouville sense) require an infinite number of independent conserved quantities. Higher dimensional systems of partial differential equations rarely possess this property. In the next section we shall describe a different approach to complete integrability that is strong enough to overcome these difficulties.

Section 1.4   New Ideas on Complete Integrability for Equilibrium Processes

Equilibrium processes involve elliptic partial differential equations, and their associated boundary value problems. Complete integrability for such problems is generally not considered. Here we show, with a simple yet definitive example, that the point of view discussed in the sections above can accommodate our expanded ideas.

The Nonlinear Boundary Value Problem

For explicitness we shall study the following nonlinear elliptic Dirichlet problem:

$$(2) \qquad \begin{cases} \Delta u + f(u) = g \\ \quad u|_{\partial\Omega} = 0 \end{cases}$$

Here Ω is a bounded domain in $\mathfrak{R}^N$ with boundary $\partial\Omega$ and $f(u)$ is a C^k convex function of u ($k \geq 2$), with the asymptotic behavior

$$(3) \qquad 0 < \lim_{t\to-\infty} f'(t) < \lambda_1 < \lim_{t\to-\infty} f'(t) < \lambda_2$$

where λ_1 and λ_2 denote the lowest two eigenvalues of the Laplace operator on Ω relative to null Dirichlet boundary conditions.

Ambrosetti and Prodi showed that for smooth $g \in C^{0,\alpha}(\Omega)$ the number of solutions of (a) is either 0, 1 or 2. In fact we shall show the notion of complete integrability described above ,using global normal forms, is applicable and show that (2) is completely integrable in this sense, <u>independent of the domain Ω and the</u>

dimension N. Moreover, we show the methods we use to establish the complete integrability of (2) are "stable" under perturbation in the sense that they yield results on the perturbed problem. In fact for a sufficiently small smooth perturbation the global normal form remains invariant exactly as in the linear case with the rank normal form.

The Notion of Complete Integrability

We now redefine our notion of complete integrability for the problem (2), in accord with our discussion in the above sections.

To this end, let A denote a given smooth mapping between two Banach spaces X_1, X_2. Then we say A is C^k equivalent to a mapping B if there are C^k diffeomorphisms α and β such that the following diagram commutes

$$
\begin{array}{ccc}
 & A & \\
X_1 & \rightarrow & X_2 \\
\alpha\uparrow & & \uparrow\beta \\
X_1 & \rightarrow & X_2 \\
 & B &
\end{array}
$$

(4)

i.e $A\alpha(x) = \beta B(x)$ for each $x \in X_1$. This just means that the mappings A and B differ by smooth coordinate changes. (Of course, we could extend this idea by assuming B is a mapping between two other Banach spaces Y_1 and Y_2 provided the diffeomorphisms α and β are defined appropriately).

Now, since we want to preserve the "Hamiltonian nature" of the problem (2) we shall choose $X_1 = X_2$ as separable Hilbert spaces. Then A can be regarded (relative to an orthonormal basis $(x_1, x_2, \ldots)$) as the coordinate mapping

(5) $A(x_1, x_2, x_3 \ldots) = (A_1, A_2, A_3 \ldots)$

The mapping A will be called C^k <u>completely</u> <u>integrable</u> if there is a mapping B of the form

$$B(x_1, x_2, x_3, \ldots) = (g_1(x_1), g_2(x_2), g_3(x_3), \ldots)$$

such that A and B are C^k equivalent $(k \geq 0)$.

To relate this notion to the boundary value problem (2), we define a mapping A between the Sobolev spaces $H = \dot{W}_{1,2}(\Omega)$ as follows:

(6) $(Au, \phi)_H = \int_\Omega \{\nabla u \cdot \nabla \phi - f(u)\phi\} dx$

for every $\phi \in H$.

Symbolically, we write

(7) $Au = \Delta u + f(u)$

where the operator A is taken in the generalized sense as a mapping of the Hilbert space H into itself.

It is easy to show

Lemma 1   A is a C^1 mapping of the Hilbert space $H = \dot{W}_{1,2}(\Omega)$ into itself.

We now state one of our main results on complete integrability.

Theorem 2   The mapping A is C^0 completely integrable in the sense that there are canonical homeomorphisms such that A is equivalent to the mapping $B : H \rightarrow H$ defined by $B(x_1, x_2, x_3, \ldots) = (x_1^2, x_2, x_3, \ldots)$.

In this result we can choose the orthonormal basis $(x_1, x_2, \ldots)$ as the normalized eigenfunctions of the Laplace operator relative to null Dirichlet boundary conditions on $\partial\Omega$.   Simple applications of this result are

Corollary 3   The mapping A defined by (7) is a proper mapping.

Proof   The mapping B of Theorem 2 is proper (i.e. the inverse image of a compact set is itself compact)and the notion of proper is preserved under C^0 equivalence.

Corollary 4   The solutions of (2) can be found explicitly (provided they exist), in terms of the "canonical coordinate changes" and the eigenfunctions of the Laplace operator.

Corollary 5   All the singular points of the mapping A are "infinite dimensional" folds in the sense of Whitney.

Thus the mapping A defined by (7) is the "simplest" nonlinear operator that is associated with a nonlinear Dirichlet problem and exhibits bifurcation phenomena <u>independent</u> of the domain Ω and the dimension N.

Additional Remark   If we suppose $f \in C^k$ $(k \geq 2)$ we may show that A is $C^{(k-2)}$ equivalent to   B provided we work with Banach spaces of Holder continuous functions.

Idea of the Proof of Theorem 2

The proof divides into two distinct parts:

Part I – An <u>analytic</u> <u>part</u> consisting of four steps:

<u>Step</u> 1: Reduction to a finite dimensional problem;
<u>Step</u> 2: Explicit cartesian representation for the singular points of A;
<u>Step</u> 3: Explicit cartesian representation for the singular values of A;
<u>Step</u> 4: Coerciveness estimates for the mapping A.

We sketch the main ideas of this part.

We write the mapping $A : H \to H$ in the form associated with the orthogonal decomposition $H = \text{Ker}(\Delta + \lambda_1) \oplus H_1$, i.e. we write an element $u \in H$ in the form $u = tu_1 + \omega$ (where u_1 is a normalized eigenfunction of Δ on Ω associated with λ_1) and so $u_1 > 0$ in Ω, with $\omega \in H_1$.   Then we show <u>for fixed</u> t that the mapping A_1 defined by

$$(7) \qquad (A_1(t, \omega), \phi) = \int_{\Omega} \left[\nabla\omega \cdot \nabla\phi - f(tu_1 + \omega)\phi \right]$$

(for $\phi \in H_1$) is a global homeomorphism of H_1 into itself. This is achieved by using the Lax-Milgram theorem to prove that the inequality

$$(A_1'(t, \omega)\phi, \phi) \geq \frac{\varepsilon}{\lambda_2} \|\phi\|_H^2$$

implies $\left\| \left[A_1'(t, \omega) \right]^{-1} \right\| \leq \frac{\lambda_2}{\varepsilon}$ for fixed $\varepsilon > 0$. The global result follows now from Hadamard's theorem on global diffeomorphisms.(cf. the references in the Bibliography) Then we find that the singular points and values can be determined by the coordinate representation

(8) $$A(tu_1 + \omega) = h(t)u_1 + g_1$$

Or more explicitly, set $u(t) = tu_1 + \omega(t)$

(9) $$\underline{A}u(t) + f(u(t)) = h(t)u_1 + g_1$$

Let us examine what happens as a singular value, so that $h'(t) = 0$.

Lemma   At a singular value with $t = t_0$

(10) $$h''(t_0) = \int_\Omega f''(u(t)) [u'(t_0)]^3$$

so that by our assumptions $h''(t_0) > 0$.

Sketch of Proof   Consider (9) and differentiate twice with respect to t assuming $h'(t_0) = 0$.

(11) $$\Delta u'(t) + f'(u(t))u'(t) = h'(t)u_1$$

Since $h'(t_0) = 0$, $u'(t_0)$ is a nontrivial solution of (10) and by the asymptotic conditions (3) we may suppose $u'(t_0) > 0$ in Ω (see [2]).

(12) $$\Delta u''(t) + f'(u(t))u''(t) + f''(u(t))[u'(t)]^2 = h''(t_0)u_1$$

Since $u''(t_0)$ is a nontrivial solution of (11) for this inhomogeneous equation

(13) $$\{f''(u(t_0)) [u'(t_0)]^2 - h''(t_0)u_1\} \perp \text{Ker} [\Delta + f'(u(t_0)]$$

$$\text{i.e.,} \int_\Omega [f''(u(t_0)) [u'(t_0)]^2 - h''(t_0)u_1] u'(t_0) = 0$$

This relation shows (10). This result and the convexity of $f(u)$ yield the lemma.

Another important fact is that $h(t) \to \infty$ as $t \to \infty$. This fact follows from the representation

$$h(t) = -\lambda_1 t + \int_\Omega f(tu_1 + \omega(t))u_1$$

the asymptotic relation (3) and the fact that as $t \to \infty$ the contribution due to $\omega(t)$ is negligible via the a priori estimate

$$\| \omega'(t,g_1) \|_H \leq c \quad \text{(independent of } t \text{ and } g_1)$$

The following picture illustrates the behavior of the function $h(t)$.

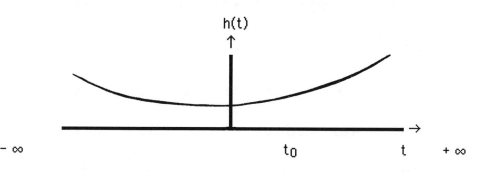

From this picture we read off the Cartesian representation of the singular points and singular values of A.

Part II. The second part of the proof is geometric, namely construction of the diffeomorphisms α and β using the fact of Part I.

This part consists of four steps also.

Step 1: Layering of the mapping A in accord with Step I of Part I by a diffeomorphism α_1.

Step 2: "Translation" of the Singular Points of the Mapping A to those of B by a diffeomorphism α_2;

Step 3: Translation of the Singular Value of A to those of B by a diffeomorphism α_3;

Step 4: The final homeomorphism.

Indeed, after Step 3 we find

(14) $\alpha_3 A \alpha_1 \alpha_2 = (\alpha(t, \omega), \omega)$

Using Step 4 of Part I we represent the right hand side of (14) by the composition $B\varphi$ where φ is a diffeomorphism $H \rightarrow H$.

Thus

(15) $\alpha_3 A \alpha_1 \alpha_2 = B\varphi$

which is the desired equation.

Stability of the Methods used under Perturbation of A

Under a C^1 perturbation of A in the sense of the metric in H, our analytical results of Step 1 carry over to study perturbation problems. Indeed, we prove

Theorem   Under a suitably restricted C^1 perturbation of A, the number of solutions of the perturbed problem is <u>exactly</u> <u>the</u> <u>same</u> as in (2) away from a neighborhood of the singular values. Moreover, in this case, the solutions of (2) are <u>accurate</u> <u>approximations</u> to the perturbed problem.

The proof is based on a careful analysis of the steps in Part I above.

A) Background and Mathematical Details

Here we begin by considering the nonlinear Dirichlet problem (2) satisfying the asymptotic conditions (3). Note that with these conditions $f''(t)$ is uniformly bounded. By standard facts λ_1 is a simple eigenvalue of Δ with the normalized eigenfunction $u_1 > 0$ in Ω and $\|u_1\|_{L_2} = 1$.

It was shown that (2) can be reformulated as an operator equation in the Sobolev space $\mathring{W}_{1,2}(\Omega)$. Recall that $\mathring{W}_{1,2}(\Omega)$ is the Hilbert space consisting of all functions $u \in L_2(\Omega)$ with gradients $|\nabla u| \in L_2(\Omega)$ and can be obtained as limits of $C_0^\infty(\Omega)$ functions relative to the norm

$$\|u\|_H^2 = \int_\Omega (u^2 + |\nabla u|^2)$$

This operator equation can be written

$$Au = \tilde{g}$$

where the operator $A: \mathring{W}_{1,2}(\Omega) \to \mathring{W}_{1,2}(\Omega)$ is defined implicitly by the formula

(3')
$$(Au, \phi) = \int_\Omega [\nabla u \cdot \nabla \phi - f(u)\phi]$$

for arbitrary $u, \phi \in \mathring{W}_{1,2}(\Omega)$. Moreover the following facts can be proven concerning the mapping A of $H = \mathring{W}_{1,2}(\Omega)$ into itself.

1(i). Differentiability property   A is a C' mapping of H into itself with A'(u)v   defined implicitly by the formula for all $\phi \in \dot{W}_{1,2}(\Omega)$

$$(A'(u)v, \phi) = \int_{\Omega} [\nabla v \cdot \nabla \phi - f'(u)v\phi]$$

Moreover, it easily follows from standard   arguments of partial differential equations that A'(u) is a compact perturbation of the identity for each $u \in H$. Thus A is a nonlinear Fredholm operator of index zero.

1(ii). Layering property   Let $H = \text{Ker}(\Delta + \lambda_1) \oplus H_1$ and P be the orthogonal projection of H onto H_1. Then an arbitrary element $u \in H$, relative to this decomposition can be written $u = tu_1 + \omega$ where $\omega \in H_1$. Set $A_1 = (I - P)A$ and $A_2 = PA$ so the mapping A can be written

$$A = (A_1(t, \omega), A_2(t, \omega))$$

In [2], it is proven that for fixed t, the mapping $A_2(t, \omega): H_1 \rightarrow H_1$ is a diffeomorphism with inverse denoted $\omega(t, g_1)$ where $g_1 = A_2(t, \omega)$, so that $\omega(t, g_1)$ is a C' function of g_1 for fixed t. Moreover, the following *a priori* bounds hold for the partial derivatives of $\omega(t, g)$

$$\|\omega_t\|_H \leq c_1, \quad \|\omega_g\| \leq c_2,$$

where c_1 and c_2 are constants independent of t and g.

1(iii). Description of the singular points and values of A and the range of A   The layering property (ii) is used to describe Cartesian representations for the singular points and singular values of A. The mapping $A_1(t, \omega) = (I - P)A$ is denoted $h(t, g_1)$, and the equation

$$Au = g$$

can be rewritten

$$A(tu_1 + \omega(t, g_1)) = h(t, g_1)u_1 + g_1$$

Setting $u(t) = tu_1 + \omega(t, g_1)$ and differentiating this equation with respect to t we find

$$A'(u(t))u'(t) = h'(t, g_1)u_1$$

It is proven that (a) the singular points of A are those points $u(t)$ where $h'(t, g_1) = 0$, (b) $h(t, g_1)$ is a C' function of t and for fixed g_1, $\lim_{|t| \to \infty} h(t, g_1) = \infty$ and moreover (c) $h'(t, g)$ vanishes exactly once at $t = t_1(g)$ where $\min_t h(t, g_1) = h(t_1, (g))$. Thus (d) the singular points of A can be written $u(t_1) = t_1(g_1)u_1 + \omega(t_1, g_1)$ and also that $\operatorname{sgn} u'(t_1) > 0$ in Ω. Moreover the singular values can be written

$$g = h_1(t_1(g_1)) + g_1$$

$$= \alpha(g_1)u_1 + g_1$$

for fixed $g_1 \in H_1$.

Consequently (f) the range of the mapping A consists of those points $g = h(t)u_1 + g_1$ where $h(t) \geq \min_t h(t, g_1)$ and (g) a singular value of A has exactly one inverse image under A, while any nonsingular value in the range of A has exactly two inverse images.

1(iv). Additional properties of the mapping A   In order to establish our integrability and stability results, we shall need to prove some additional properties of the mapping A based on (i) – (iii) above. In fact, using the notation introduced above, we shall prove

(a) $\min_t h(t, g_1)$ depends continuously on g_1.

(b) If the real number $t_1(g_1)$ is defined by setting $h(t_1(g_1)) = \min_t h(t, g)$, then $t_1(g_1)$ depends continuously on g_1.

(c) The mapping A is proper.

(d) At a singular point of A, $h(t)$ is twice differentiable and $h''(t) > 0$.

We begin with the following:

Lemma   If $g_n \rightarrow g$ in $L_2(\Omega)$, then $\omega(t, g_n) \rightarrow \omega(t, g)$ uniformly in t.

Proof   By definition, $\omega = \omega(t, g)$ is defined implicitly by equation

$$\int_\Omega [\nabla\omega \cdot \nabla\phi - f(u(t))\phi] = \int_\Omega g\phi, \quad \forall\phi \in H_1$$

Setting $W = \omega(t, g) - \omega(t, g_n)$ and $u_n(t) = tu_1 + \omega(t, g_n)$ we find from this equation

$$\int_\Omega \nabla W \cdot \nabla \phi = \int_\Omega [f(u(t)) - f(u_n(t))]\phi + \int_\Omega (g_n - g)\phi$$

Letting ϕ carry over the unit sphere in H_1 and using the fact that $\|\phi\|_{L_2} \leq \lambda_2^{-1/2} \|\phi\|_{H_1}$, we find from the Cauchy-Schwarz inequality

$$\|W\|_{H_1} \leq \lambda_2^{-1/2}(\lambda_2 - \varepsilon_+) \|W\|_{L_2} + \lambda_2^{-1/2} \|g_n - g\|_{L_2}$$

Here we have used the fact that

$$|f'(t)| \leq \lambda_2 - \varepsilon_+$$

Since $W \in H_1$ we have from the above

$$\|W\|_{H_1} \leq (\lambda_2 - \varepsilon_+)\lambda_2^{-1}\|W\| + \lambda_2^{-1/2}\|g_n - g\|_{L_2}$$

Consequently, we find

$$\|W\|_{H_1} \leq \frac{\lambda_2^{1/2}}{\varepsilon_+} \|g_n - g\|_{L_2}$$

This is the required uniform bound.

Lemma $\min_t h(t, g)$ depends continuously on $g \in L_2$.

Proof   Let $\zeta(g) = \min_t h(t, g)$, where

$$h(t, g) = -\lambda_1 t + \int_\Omega [f(tu_1 + \omega(t, g))]u_1 ,$$

and let $g_n \to g$ in L_2. Now set

$$\zeta(g) = \sup_t \left\{ \lambda_1 t + \int_\Omega [f(tu_1 + \omega(t, g))]u_1 \right\}$$

$$= \sup_t \{-h(t, g)\}$$

Thus since $\sup A - \sup B \le \sup(A - B)$,

$$\zeta(g_n) - \zeta(g) \le \sup_t \int_\Omega u_1 \{f(tu_1 + \omega(t, g_n)) - f(tu_1 + \omega(t, g))\}$$

Since

$$|f(s_1) - f(s_2)| \le \sup_s f'(s)|s_1 - s_2|$$

$$\le (\lambda_2 - \varepsilon_+)|s_1 - s_2|,$$

from the Cauchy-Schwarz inequality and $\|u_1\|_{L_2} = 1$,

$$\zeta(g_n) - \zeta(g) \le (\lambda_2 - \varepsilon_+)\sup_t \|\omega(t, g_n) - \omega(t, g)\|_{H_1}$$

Since we may interchange g_n and g, $|\zeta(g_n) - \zeta(g)|$ has the same bound, and the conclusion results from the Lemma above.

Theorem   The mapping A defined by (3') is a proper mapping of $W_{1,2}(\Omega)$ into itself.

Proof   By virtue of a well known result (see Berger [Nonlinearity and Functional Analysis, page 102]) it suffices to prove that point inverses of A are compact and in addition that A is a closed mapping. The fact that point inverses are compact (in this case even finite sets) follows of course from the properties (iii) described above, and indeed these properties show in addition that the mapping A is closed. Let $C \subset \dot{W}_{1,2}(\Omega)$ be a closed set, we show $A(C)$ is closed in $\dot{W}_{1,2}(\Omega)$.

To this end, let $\tilde{g}_n \to \tilde{g}$ in H, $\tilde{g}_n \in A(C)$. Then to show $\tilde{g} \in A(C)$, we note that by (iii) it suffices to write

$$\tilde{g} = \alpha u_1 + g_1 \qquad \text{with } g_1 \in H_1$$

and to show that $\alpha \geq \min_t h(t, g)$. To this end, since $\tilde{g}_n \in A(C) \tilde{g}_n = \alpha_n u_1 + g_n$ has the property that

$$(*) \qquad\qquad \alpha_n \geq \min_t h(t, g_n)$$

Since $\tilde{g}_n \to \tilde{g}$, $\alpha_n \to \alpha$, and by the continuity property expressed by Lemmas above,

$$\min_t h(t, g_n) \to \min_t h(t, g)$$

Consequently, letting $n \to \infty$ in the inequality $(*)$ we find $\alpha \geq \min_t h(t, g)$, and the properness of A is established.

Lemma $t_1(g)$ depends continuously on g, where $t_1(g)$ is the unique t such that $\min_t h(t, g)$ is attained.

Proof Suppose $g_n \to g$ in H_1. Since

$$\min_t h(t, g) = h(t_1(g), g),$$

The Lemma above implies that

(**) $h(t_1(g_n), g_n) \to h(t_1(g), g)$

Now $A(t_1(g_n)u_1 + \omega(t_1(g_n), g_n)) = h(t_1(g_n), g_n)u_1 + g_n$, and it follows from the properness of A that the sequence

$$\{t_1(g_n)u_1 + \omega(t_1(g_n), g_n)\}$$

has a convergent subsequence, i.e. relabelling

$$t_1(g_n)u_1 + \omega(t_1(g_n), g_n) \to t_* u_1 + \omega \qquad \text{(say)}.$$

From (**) and the continuity of A,

$$A(t_* u_1 + \omega) = h(t_1(g), g)u_1 + g,$$

so $h(t_*, g) = h(t_1(g), g)$. Since the minimum is attained at a unique point, $t_* = t_1(g)$.   Our argument shows that $\{t_1(g_n)\}$ has a subsequence convergent to $t_* = t_1(g)$. Since $\{g_n\}$ was an arbitrary sequence with $g_n \to g$, t_1 is continuous.

Theorem   Let $u(t_0) = t_0 u_1 + \omega(t_0, g_1)$ be a singular point of the mapping A, then $h(t)$ is twice differentiable at t_0, and moreover

$$h''(t_0) = \int_\Omega f''(u(t_0))[u'(t_0)]^3$$

Proof   We utilize   a formula concerning h(t)   near a singular point, namely at the point $t = t_0$

$$h'(t) - h'(t_0) = \int_{\Omega} [f'(u(t)) - f'(u(t_0))]u'(t)u'(t_0)$$

then assuming that the limit can be taken under the integral sign we find

$$h''(t_0) = \lim_{t \to t_0} \left[\frac{h'(t) - h'(t_0)}{t - t_0} \right] = \int_{\Omega} f''(u(t_0))[u'(t_0)]^3$$

Now the convexity of f and the fact that at a singular value $u'(t_0) > 0$ in Ω imply $h''(t_0) > 0$. It remains to justify the passage of the limit under the limit sign. To this end we use the Lebesgue dominated convergence theorem and thus must demonstrate that: (a) $f''(u(t_0))[u'(t_0)]^3$ is integrable and (b) $\left[\dfrac{f'(u(t)) - f'(u(t_0))}{t - t_0} \right] u'(t)u'(t_0)$ is bounded by an integrable function. To prove (a) we note that since $f''(t)$ is uniformly bounded, it suffices to prove that $[u'(t_0)]^3$ is integrable over Ω, i.e. $u'(t_0) \in L_3(\Omega)$. Now $u'(t_0)$ is a weak solution on Ω in $\dot{W}_{1,2}(\Omega)$ of the equation

$$\Delta u'(t) + f'(u(t))u'(t) = 0$$

with $f'(u(t))$ bounded and measurable. Consequently, by the regularity theory for weak solutions of linear elliptic equations, $u'(t) \in W_{2,p}(\Omega)$ for arbitrarily large but finite p. Consequently, $u'(f) \in L_3(\Omega)$ by Sobolev's inequality. In fact, $u'(t_0)$ is continuous.

To prove (b), we note that

$$\left|\frac{f'(u(t)) - f'(u(t_0))}{t - t_0}\right| \leq \sup_s |f''(s)| \left[\left|\frac{u(t) - u(t_0)}{t - t_0}\right|\right]$$

Since $f''(s)$ is uniformly bounded, the desired result will be established once we show $u'(t)u'(t_0)\left[\dfrac{u(t) - u(t_0)}{t - t_0}\right]$ is integrable. But since $u'(t_0)$ is continuous and hence bounded as mentioned above and $u'(t) \in L_2$ we find as required

$$\int_\Omega \left|u'(t)u'(t_0)\left[\frac{u(t) - u(t_0)}{t - t_0}\right]\right| \leq \sup|u'(t_0)| \; \|u'\|_{L_2}^2$$

Fact　　The operator A defined above is a C^2 mapping of H into itself provided the function $f(t)$ is C^3, has a uniformly bounded third derivative and $\dim \Omega \leq 4$. Also $A''(u)$ is defined by the following formula

$$(A''(u)v_1 v_2, \phi) = -\int_\Omega f''(u)v_1 v_2 \phi \qquad \forall \, \phi \in H$$

Proof　　By virtue of the formulae for $A'(u)$ given above, we find the second Gateaux derivative of A by computing

$$\frac{d}{d\varepsilon}(A'(u + \varepsilon v_2)v_1, \phi)\big|_{\varepsilon=0} = -\frac{d}{d\varepsilon}\int_\Omega f'(u + \varepsilon v_2)v_1 \phi\big|_{\varepsilon=0}$$

Thus

$$(A''(u)v_1 v_2, \phi) = -\int_\Omega f''(u)v_1 v_2 \phi$$

The passage of the differentiation under the integral sign being justified provided the integrand in the last integral is Lebesgue integrable. Using the fact that $\dim \Omega \leq 4$ we find by Sobolev's inequality that $v_1, v_2 \in L_4(\Omega)$; consequently the desired integrability follows from the multiple Hölder inequality. Indeed since $f''(s)$ is uniformly bounded by C (say)

$$\int |f''(u)v_1 v_2 \phi| \leq C\|v_1\|_{L_4}\|v_2\|_{L_4}\|\phi\|_{L_2} < \infty$$

To prove that A is actually Fréchet differentiable and C^2, use a standard result and compute

$$\|[A''(u) - A''(w)]v_1 v_2\| = \sup_{\|\phi\| \leq 1} \int [f''(w) - f''(u)]v_1 v_2 \phi$$

Again by Sobolev's inequality v_1, v_2, ϕ and $(w - u) \in L_4(\Omega)$, since $\dim \Omega \leq 4$. Thus by the multiple Hölder inequality

$$\|[A''(u) - A''(w)]v_1 v_2\| \leq \sup_{\|\phi\|_H \leq 1} \|f''(w) - f''(v)\|_{L_4}\|v_1\|_{L_4}\|\phi\|_{L_4}\|v_2\|_{L_4}$$

Now for any $u \in H$, $\|u\|_{L_4} \leq C\|u\|_H$, where C is an absolute constant depending only on Ω. Thus

$$\|(A''(u) - A''(w))v_1 v_2\| \leq C_3\|f''(u) - f''(v)\|_{L_4}\|v_1\|_H\|v_2\|_H$$

To estimate this last norm on the right-hand side, we note that since $f'''(s)$ is uniformly bounded by a constant K (say) by hypothesis

$$|f''(u) - f''(v)| \leq K|u - v|$$

Thus $A''(u)$ is (uniformly) continuous in u, since

$$\|[A''(u) - A''(v)]v_1 v_2\| \leq KC^4 \|u - v\|_H \|v_1\|_H \|v_2\|_H$$

Moreover, at a singular point $u(t_0)$ of A

$$h''(t_0) = (A''(u(t_0)(u'(t_0), u'(t_0)), u'(t_0))$$

Corollary   For the map A, each singular point is a fold point.

Proof   Note that by the results mentioned earlier, A is Fredholm of index 0 with $\dim \ker A'(x) \leq 1$ for every x, so Property (1) results. For (2), note that for $u(t_0) \in S$.

$$u'(t_0) = u_1 = \omega'(t_0) \in \ker A'(u(t_0))$$

and by standard results for second order elliptic partial differential equations, $u'(t_0) > 0$ in Ω. Now we prove the inner product

$$(3) \qquad \left\langle A''(u(t_0))(u'(t_0), u'(t_0) \right\rangle = \int_\Omega f'''(u(t_0))(u'(t_0))^3 > 0$$

so that $A''(u(t_0))(u'(t_0, u'(t_0)) \neq 0$ and (2) follows. Indeed, since f is strictly convex at zero, $f''(u(t_0)) > 0$ in Ω, the inequality on the right follows.

On the other hand, to demonstrate the equality in (3) above, we proceed by differentiating the following identity at the singular point $u(t_0)$

$$A(u(t)) = h(t)u_1 + g_1$$

Differentiating once with respect to t,

$$A'(u(t))u'(t) = h'(t)u_1$$

Differentiation again at $t = t_0$,

$$A''(u(t_0))(u'(t_0))^2 + A'(u(t_0))u''(t_0) = h''(t_0)u_1$$

Taking the inner product of both sides with $u'(t_0) = u_1 + \omega'(t_0)$ and using the self-adjointness of $A'(u(t_0))$ to insure that $(A'(u(t_0))u''(t_0), u'(t_0)) = 0$, we find

$$(A''(u(t_0))(u'(t_0), u'(t_0)). u'(t_0)) = h''(t_0)$$

Thus the equality in equation (3) above follows from the Theorem stated above.

Section 1.5   Canonical Changes of Coordinates for the Mapping A

In this section we show that the above results enable us to find invertible coordinate transformations that transform the mapping A defined by (3') into the mapping that defines a global Whitney fold. This is accomplished in two cases: first, homeomorphic coordinate changes, and later (assuming A is C^2), diffeomorphically with the same coordinate changes as used in the first case. We can be shown that this complete integrability leads to explicit formulae for the solutions of (2a) and (2b).

We begin with some definitions.

Definition   Two maps $f: K \to L$ and $g: M \to N$ are C^k-equivalent if there are C^k diffeomorphism $\varphi: K \to M$ and $\psi: L \to N$ such that the following diagram commutes

$$
\begin{array}{ccc}
K & \overset{\varphi}{\to} & M \\
f \downarrow & & \downarrow g \\
L & \underset{\psi}{\to} & N
\end{array}
$$

Here a C^0 map is a continuous function, a C^0 diffeomorphism, and if $k \geq 1$, the spaces are assumed to be C^k manifolds.

Definition   For a $C^k(k \geq 1)$ map $A: H \to H$ the singular set $S = S(A)$ will be the set of $x \in H$ with $A'(x) = 0$; by the Inverse Function theorem, if $x \notin S(A)$, then A is locally a C^k diffeomorphism of a neighborhood of $A(x)$. If A is C^0 (i.e. continuous), we define the topological singular set $S'(A)$ by :

$x \notin S'(A)$ if and only if A is locally a homeomorphism of a neighborhood of x onto a neighborhood of $A(x)$. If A is C^k $(k \geq 1)$, then $S'(A) \subset S(A)$.

Coordinate Changes

Theorem   The map A defined by equation (3') is C^0 equivalent to $F: \mathcal{R} \times H_1$ defined by $F(t, \omega) = (t^2, \omega)$ where H_1 is the separable Hilbert space orthogonal to the vector $u_1 \in \dot{W}_{1,2}(\Omega)$.

Proof   Write $A: \mathcal{R} \times H_1 \to \mathcal{R} \times H_1$ as $A(t, \omega) = (A_1(t, \omega), A_2(t,\omega))$, as described above. We will apply in succession four coordinate changes so that A is transformed to maps with successively better properties:

(i) first, to be the identity in the second coordinate;

(ii) second, to have topological singular set (contained in) $0 \times H_1$, as well,

(iii) third, to have topological singular set image (contained in) $0 \times H_1$, as well,

(iv) and last, to be the desired map F.

For each fixed t by the results summarized in Section 1 the map $\omega \to A_2(t, \omega)$ is a diffeomorphism C^1 since A is C^1, so

$$\Gamma: \mathcal{R} \times H_1 \to \mathcal{R} \times H_1, \qquad (t, \omega) \to (t, A_2(t, \omega))$$

is a diffeomorphism;   let Λ be its inverse.   Then $\Lambda(t, \omega) = (t, \Lambda_2(t, \omega))$, and

(1) $A\Lambda(t, \omega) = (A_1(t, \Lambda_2(t, \omega)), \omega)$,

(2) the singular set $S(A\Lambda)$ is the set of (t, ω) with

$$\frac{\partial}{\partial t}(A_1(t, A_2(t, \omega))) = 0$$

and $A\Delta$ is C^1-equivalent to A.

As noted above, for each fixed $t \in \mathcal{R}, \omega \to A_2(t, \omega)$ is a diffeomorphism. Above, its inverse is denoted by $g_1 \to \omega(t, g_1)$, so $\omega(t, g_1)$ is our $A_2(t, g_1)$. The function

$$h(t, g_1) = A_1(t, A_2(t, g_1)) = \pi_1(A\Delta(t, g_1)),$$

where $\pi_1 = R \times H_1 \to \mathcal{R}$ is projection. For each fixed $g_1 \in H_1$ there is a unique t, called $t_1 = t_1(g_1)$, such that $h'(t_1, (g_1)) = 0$. Thus (if we replace g_1 by ω) the singular set

(3) $S(A\Delta) = \{(t_1(\omega), \omega): \omega \in H_1\}$

by (2). Moreover, by the above discussion of Section 1.4, $t_1: H_1 \to \mathcal{R}$ is continuous.

The maps $\theta_i : \mathcal{R} \times H_1 \to \mathcal{R} \times H_1$ $(i = 0, 1)$ defined by

$$(t, \omega) \to (t + (-1)^i t_1(\omega), \omega)$$

are continuous, and since each is the inverse of the other, each is a homeomorphism. Thus $A\Delta\theta_0$ is C^0-equivalent to A, with topological singular set $S'(A\Delta_0) = \theta_1(S'(A\Delta)) \subset 0 \times H_1$.

The map $\sigma: H_1 \to \mathcal{R}$ defined by

$$\omega \to A_1(t_1(\omega), A_2(t_1(\omega), \omega)) = h(h(t_1(\omega))$$

is continuous, and by (1) and (3) the singular set image

$$A(S(A)) = A\Lambda(S(A\Lambda)) = \{A\Lambda(t_1(\omega), \omega): \omega \in H_1$$

$$= \{(\sigma(\omega), \omega): \omega \in H_1\} .$$

Define $\psi: \mathcal{R} \times H_1 \to \mathcal{R} \times H_1$ by

$$\psi(t, \omega) = (t - \sigma(\omega), \omega)$$

As above Ψ is a homeomorphism with $\Psi(A(S'(A))) = 0 \times H_1$, since $h''(t) > 0$ at each singular point by the Theorem above, $\Psi(\text{imag } A) = [0, \infty) \times H_1$ rather than $(-\infty, 0] \times H_1$.

Now $\Psi A\Lambda\theta_0$ is C^0-equivalent to A, $\Psi A\Lambda\theta_0(t, \omega)$ is in the form $(\alpha(t, \omega), \omega)$, and its topological singular set and topological singular set image are both $0 \times H_1$. According to the results above, for each fixed $g_1 \in H, h(t) \to +\infty$ as $|t| \to \infty$, and h has only one critical point, at $t_1(g_1)$. Thus h maps $[t_1(g_1), \infty)$ and $(-\infty, t_1(g_1)]$ each homeomorphically onto $[h(t_1(g_1)), \infty)$. It follows that for each fixed $\omega \in H_1$.

(4) α maps $[0, \infty)$ and $(-\infty, 0]$ each homeomorphically onto $[0, \infty)$, and $\Psi A\Lambda\theta_0$ maps $[0, \infty) \times H_1$ and $(-\infty, 0) \times H_1$ each homeomorphically onto $[0, \infty) \times H_1$. Thus $\alpha(0, \omega) = 0$, and $\alpha(t, \omega) > 0$ for $t \neq 0$.

Define $\gamma(t, \omega) = [\alpha(t, \omega)]^{1/2}$ for $t \geq 0$ and $= -[\alpha(t, \omega)]^{1/2}$ for $t \leq 0$, and define $Y: \Re \times H_1 \to \Re \times H_1$ by $Y(t, \omega) = (\gamma(t, \omega), \omega)$. From (4) if follows that Y is a homeomorphism mapping $[0, \infty) \times H_1$ onto itself and $(-\infty, 0) \times H_1$ onto itself. Define $\Phi = Y\theta_1\Gamma$ so Φ and Ψ are homeomorphism as desired. Remember that $F(t, \omega) = (t^2, \omega)$. Now $FY(t, \omega) = ((\gamma(t, \omega))^2, \omega) = \alpha(t, \omega), \omega) = \Psi A \Lambda \theta_0(t, \omega)$; thus $\Psi A = FY(\Lambda \theta_0)^{-1} = FY\theta_1\Gamma = F\Phi$, so $\Psi A = F\Phi$, as desired.

Section 1.6   Bifurcation and the Integration of Nonlinear Ordinary and Partial Differential Equations

Can integrability be found when bifurcation is present?

This question, by its very nature, requires a mathematical analysis; first, of the notion of integrability and secondly, of the bifurcation process.

Intuitively, we are dealing with two extreme situations in nonlinear dynamics. First, there is the notion of a completely integrable Hamiltonian dynamic system. On the other hand, many nonlinear dynamical systems exhibit the reverse of integrability, namely "chaotic" behavior, by which we mean sensitivity to perturbations of the initial conditions, as well as unpredictable asymptotic behavior. Our goal in this work is to find a case intermediate between these two extremes that incorporates bifurcation phenomenon as an intrinsic part of the integrability idea.

The significance of our extension is the realization that the integrability idea is a global one whereas bifurcation and chaotic behavior is formulated form a local point of view. Thus we must attempt to understand bifurcation phenomenon more globally in order to address our fundamental key problem. This will be one of the subjects in this section.

Extension of Fold Singularities to Nonlinear Evolution Operators

Motivation   In the above discussion two key examples of the new integrability idea were discussed:

Example 1   The periodic Riccati operator involving a single time variable t.

Example 2   The spatial nonlinear Dirichlet problem defined over Ω, an arbitrary bounded domain.

        To extend our ideas to nonlinear evolution equations we combine the time periodicity of Example 1 with the nonlinearity and the Dirichlet spatial behavior of Example 2.

Nonlinear Parabolic Evolution Equations

        We consider the equation

(*)
$$\frac{du}{dt} + Lu + f(u) = g$$

$$u|_{\partial\Omega} = 0$$

subject to Dirichlet data on the boundary of a bounded domain Ω in $\mathcal{R}^N$ and L is a strongly elliptic second-order linear differential operator. The conventional theory for the equations of this class consists of studying the Cauchy problem for (*). This gives rise to the theory of semi-groups and has many successes.   However, currently one runs into difficulties with this approach because of chaotic and bifurcation phenomenon. Moreover, the methods of inverse scattering for this problem seem to break down when the space dimension $N > 1$ and when a given nonlinear problem integrable by inverse scattering is slightly perturbed. To overcome this difficulty we shall change the Cauchy problem and consider periodic solutions of the associated parabolic system. This in effect means that we add boundary conditions in the T-variable and this will allow us to study bifurcation phenomenon from a more conventional point of view.

Thus we add to (*) the periodic boundary condition reflecting the T-periodicity in the function g on the right hand side of (*) so that we supplement (*) by the boundary condition

(**) $u(0) = u(T)$

The next step in our set-up is to translate the problem (*) – (**) into a functional analysis setting and to apply the methods of the previous sections and the discussion thereafter. Thus we attempt to define Hilbert spaces X and Y so that the operator $A: X \rightarrow Y$ defined by

$$Au = \frac{du}{dt} + Lu + f(u)$$

subject to the appropriate boundary conditions is a nonlinear operator of Fredholm index zero between the spaces of X and Y. In fact, we have the following result:

Theorem   Choose $X = W_{1,2}(0, T; H)$ and $Y = L_2(0, T; H)$ with $H = \dot{W}_{1,2}(\Omega)$. Then the operator $A(u)$ defined in the equation above is a nonlinear Fredholm operator of index zero acting between the spaces of X and Y.

The next step in our analysis is to analyze the singular points of the mapping A, to investigate the bifurcation points and bifurcation values of this mapping and to attempt to classify these bifurcation points as folds.

Consider the nonlinear operator

(***) $Bu = u_t + Lu + f(x, u)$

acting between function spaces X and Y consisting of functions u that are T-periodic in t (the time variable). The functions u are assumed to be defined on an arbitrary bounded domain Ω in $\mathbb{R}^N$ but are restricted to satisfy null Dirichlet boundary conditions on the boundary of Ω, $\partial\Omega$.

Here we assume L is a second order elliptic operator, like the Laplace operator Δ or simple extensions, and $f(x, u)$ satisfies the simple asymptotic relations defined by (*).

<u>Question</u> : Can the basic idea of fold singularities be extended to this more general context? <u>Answer</u> :   Yes!

To carry out the analytic details of this statement we form the simple abstract route outlined in the earlier parts of the chapter.

<u>Step</u> 1   Choose the function spaces mentioned above

$$X = W_{1,2}[(0, T), H]$$

$$Y = L_2 [(0, T), H]$$

consisting of T-periodic functions of t with $H = \dot{W}_{1,2}(\Omega)$.

<u>Step</u> 2   Recall the result stated above:

Theorem α The mapping B: X $\rightarrow$ Y defined by (***) is a nonlinear Fredholm operator of index 0.

        We sketch the proof of this result in Appendix 2 at the end of this Chapter.

<u>Step</u> 3   Consider the associated elliptic operator

$$Au = Lu + f(x, u)$$

defined on Ω satisfying the same null spatial, Dirichlet, boundary conditions on $\partial\Omega$. Regard A is a C' mapping between $\dot{W}_{1,2}(\Omega) = H$ into itself (defined by duality as described in the Section 1.4).

        We state

Theorem β For any u $\in$ H

$$\dim \text{Ker } A'(u) = \dim \text{Ker } B'(u)$$

Thus regular points of A are regular points of B. Similarly, singular points of A are singular points of B. Since regular points of A have dim Ker A'(u) = 0, whereas singular points of A have dim Ker A'(u) > 0.

<u>Step</u> 4   Show any u ∈ H that is a singular point of A is necessarily a (Whitney) fold for B, regarded as a mapping between X and Y.

This result follows immediately from the above discussion of the elliptic case and the abstract characterization of a Whitney fold also discussed above.   More details of this work on nonlinear evolution equations is carried out in Appendix 2 at the end of the chapter.

The result on nonlinear evolution equations stated here is restricted to behavior near equilibria, and so has local character. The full global result awaits new research.

Special Stability Properties of the New Integrability Method

The methods that we have been advocating in this book have a very special feature stemming from the mathematical structures from the bifurcation that we utilize.   Conventional ideas on complete integrability and inverse scattering techniques break down when the integrable equations involved are perturbed. However, the methods we are advocating are stable under perturbation, and provided the bifurcation points involved are folds or cusps, the global normal forms involved will not be destroyed under perturbation but rather the changes of coordinates defining the global normal forms will be slightly perturbed.

Of our examples certainly the Riccati equation of Section 1.3 and the nonlinear Dirichlet problem equation have a remarkable stability property.   Namely, its singularities consist entirely of folds, and in fact, more is true.   More explicitly, we have the following theorem:

Theorem   The global normal form is "stable" for the two examples stated above in the sense that a C^2 perturbation of either example preserves the associated global Whitney fold property.

The proof of this result follows immediately by analyzing the conditions yielding the global normal form of a Whitney fold in each case.

To illustrate how to go beyond the notion of an infinite dimensional Whitney fold and Hilbert space, we now give a few definitions of these special stable singular points, folds and cusps regarded as singularities of mappings that act between real Banach spaces.

Definition   Let A be C^k ($k \geq 2$) map germ at $\bar{u} \in E_1$ regarded as mapping between Banach spaces E_1 and E_2. Then A has a <u>fold</u> at $\bar{u}$ if:

    (0) A is Fredholm at $\bar{u}$ with index 0;

    (1) dim ker $DA(\bar{u})$ = 1 (and therefore range $DA(\bar{u})$ has
        codimension one); and

    (2) for some (and hence for any) nonzero element
        $e_0 \in$ ker $DA(\bar{u})$, $D^2A(\bar{u})(e_0, e_0) \notin$ range $DA(\bar{u})$.

For a map A a point $\bar{u}$ in its domain is called a <u>fold</u> <u>point</u> if the germ of A at $\bar{u}$ is a fold.

Definition   Let A be a C^k ($k \geq 2$) map germ at $\bar{u} \in E_1$. Then A is a <u>precusp</u> if:

    (0) A is Fredholm with index 0;

(1) dim ker $DA(\bar{u}) = 1$ (and therefore range $DA(\bar{u})$ has codimension one);

(2) for some (and hence for any) nonzero element $e_0 \in \ker DA(\bar{u})$, $D^2A(\bar{u})(e_0, e_0) \in$ range $DA(\bar{u})$; and

(3) for some $\omega \in E_1$, $D^2A(\bar{u})(e_0, \omega) \notin$ range $DA(\bar{u})$.

If $k \geq 3$ and A is a precusp satisfying

(4) $D^2(A|SA)(\bar{u})(e_0, e_0) \notin$ range $D(A|SA)(\bar{u})$ (where SA is the singular set or critical set of A), then A is a <u>cusp</u>.

If $k \geq 3$ condition (4) can be changed to

(4̃) $D^3A(\bar{u})(e_0, e_0, e_0) - 3D^2A(\bar{u})(e_0, (DA(\bar{u}))^{-1}(D^2A(\bar{u})(e_0, e_0))) \notin$ range $DA(u)$

The global Whitney cusp in infinite dimensions is of great interest in applications. It can be given the coordinates

$$(x_1, x_2, x_3, \ldots) \rightarrow (x_1 x_2 + x_1{}^3, x_2, x_3, \ldots).$$

Research on this global normal form is in progress.

Section 1.7   Qualitative Properties of Integrable Systems – Periodic and Quasiperiodic Motions of Dynamical Systems

A Hamiltonian system of n degrees of freedom that is integrable in the sense of Liouville possesses   n   independent conservation laws.   The   level   sets   are   denoted F_i = constant, i = 1, 2, ... n. These level sets determine a manifold M_c, which we assume is compact and connected.   Now we inquire what qualitative properties of this Hamiltonian system do its   solutions possess relative to this manifold M_c. An additional point of interest is the stability of these properties under a perturbation of this Hamiltonian system that preserves its Hamiltonian structure.   In symbols we write the perturbed Hamiltonian as

$$H = H_0 + \varepsilon H_1$$

If we take angle action variables   I, ϕ of the integrable system with Hamiltonian function $H_0 = H_0(I)$, the Hamiltonian equations for the unperturbed system can be written

$$\dot{I} = 0 \quad \dot{\phi} = \omega(I) \quad \text{where} \quad \omega(I) = \frac{\partial H_0}{\partial I}$$

and the perturbed system can be written

$$\dot{I} = \varepsilon g \quad \dot{\phi} = \omega(I) + \varepsilon f$$

Here the functions f   and   g   both depend on the angle action variables   I   and ϕ, but are related by the fact that the above equation determines a Hamiltonian system. The motions on M_c of the integrable system are all quasiperiodic with n frequencies. The manifold M_c itself is referred to as an invariant torus because M_c is homeomorphic to a torus and under the time evolution of the integrable system, it remains invariant. Thus it is natural to ask if the perturbed system has any of these qualitative properties.

Poincaré regarded this question concerning perturbations as the "fundamental problem of dynamics." The main observation concerning the perturbed system is that it is nonintegrable. In fact, as mentioned above, the n independent conservation laws for the perturbed system are generally destroyed. Thus it becomes an important question to analyze just what methods can be used to treat such a perturbed problem. The famous question of "small denominators" becomes a key ingredient in this issue, because the normal perturbation theory for this system involves power series in the variable ε that diverge for a dense set of frequencies of the unperturbed motion. A simple example of this "small denominators" problem will be discussed below. A key problem associated with small denominators is the problem of initial conditions that determine the motion of the perturbed system. Indeed, the troublesome small divisors are caused by initial conditions that lie in a dense set of real numbers. In the sequel, we shall illustrate methods of discussing simple nonintegrable systems that avoid initial value problems and thus avoid questions of small denominators.

We begin with the simplest such integrable system, a family of n coupled harmonic oscillators. Here the equation of motion for the couple system can be written

(1) $\ddot{x} + Ax = 0$ where A is a self-adjoint positive definite matrix with eigenvalues $\lambda_1^2, \lambda_2^2, \lambda_3^2 \ldots \lambda_n^2$

As mentioned above, this system has n distinct normal modes of period $2\pi/\lambda_1$, $2\pi/\lambda_2$, ... $2\pi/\lambda_n$. Moreover, every motion of the coupled system consists of a finite linear combination of these normal modes. By the principle of superposition for such linear systems as (1), this gives rise to the fact that a general vibration of a system is a quasi-periodic motion with n frequencies $2\pi/\lambda_1$, $2\pi/\lambda_2$, ... $2\pi/\lambda_n$.

From this example it is evident that the study of quasi-periodic motions themselves are an important element in the study of solutions of integrable systems. Indeed, by the Liouville integrability theorem stated above, it is clear that the quasiperiodicity of solutions of an integrable system (just demonstrated so easily for the simplest case of coupled harmonic oscillators) holds in general.

What happens when a system is not integrable relative to quasi-periodic motions? This question has historical roots dating back to Weierstrass and Poincaré relative to questions of celestial mechanics and the stability of the solar system. For example, consider a Hamiltonian system near an equilibrium chosen to be the origin. Suppose the Hamiltonian is analytic and expanded in a power series about the origin. Then, one interesting approach to this problem is due to Birkhoff. It involves finding a change of variables of a Hamiltonian system to a normal form. The idea is the following:

Let the Hamiltonian system of n degrees of freedom be written as

$$\dot{x}_k = H_{y_k}, \qquad \dot{y}_k = -H_{x_k}, \qquad k = 1, 2, \ldots, n$$

where H is a real analytic function of 2n variables. Then we consider changes of coordinates of the form

(*)
$$x = u(\xi, \eta) , \quad y = v(\xi, \eta)$$

which transform the Hamiltonian system into a simpler one of the form

(**)
$$\dot{\xi}_k = 2\eta_k \Gamma_k , \quad \dot{\eta}_k = -2\xi_k \Gamma_k$$

taking the function $H(x, y)$ of 2n variables into the new Hamiltonian $\Gamma = \Gamma(\Gamma_1, \Gamma_2, \ldots \Gamma_n)$ a real function of the n variables $\Gamma_i = \xi_i^2 + \eta_i^2$. The transformed system then has the form (**) above, which implies that the time derivative $d/dt(\Gamma_k) = 0$. Thus Γ_k is a positive constant, and indeed, a conserved quantity for (**). Thus the system (**) becomes linear and can be easily integrated. Thus the functions $\xi_k(t)$ and $\eta_k(t)$ can be simply written in the form $\exp(ip_k t)$ where p_k are constants. Consequently, in terms of the x and y variables, the equation (*) implies that the $x_k(t)$ and $y_k(t)$ appear as trigonometric sums of the form

$$\sum_j c_j e^{it(j, r_j)}$$

Here the vectors j and Γ_j have n components. Such a representation implies that the functions $x_k(t)$ and $y_k(t)$ are quasi-periodic in the variable t. The difficulty with this approach is that even if $H(x, y)$ is assumed analytic without linear or constant terms, and the quadratic terms in H denoted H_2 have the form

$$H_2 = \sum_{k=1}^{n} \frac{\alpha_k}{2}(x_k^2 + y_k^2)$$

with the coefficients α_k independent over the rationals, the change of coordinates u and v, even then, can only be assumed to be formal power series, i.e. the power series may not converge at all. This result was stated in Birkhoff's famous book on dynamical systems. If the formal power series converged, the result would imply that the functions Γ_k would form a set of n independent integrals in involution, and consequently the transformed system (**) would be integrable.

A Survey of Quasiperiodic Motions of Mechanical Systems

Since the days of Lagrange and Poincaré, it is customary to study periodic motions of mechanical systems with n degrees of freedom. Numerous methods have been introduced to study such motions involving variational principles, fixed point methods, asymptotic expansions, etc. However, recent developments in chaotic dynamics show the importance of going beyond such periodic behavior to consider motions of a mechanical system that are not necessarily periodic but bear many resemblances to such motions.

As mentioned above, a problem arises in all questions concerning quasiperiodic motions of nonlinear systems. Simply put the problem is this:

(A) How can one find quasiperiodic motions of arbitrary amplitude for systems that are not integrable?

To address this question in a systematic way, we consider the Fourier development of an arbitrary quasiperiodic function.

Basically the function $h(t)$ is called quasiperiodic with the n frequencies $\omega_1, \omega_2 \ldots \omega_n$ if there is a smooth function H of n variables $x_1, x_2, \ldots x_n$ such that H is 2π periodic in each of the n variables and in addition

$$h(t) = H(\omega_1 t, \omega_2 t \ldots \omega_n t)$$

The simplest case occurs when the function H depends on only one variable. Then the function $h(t)$ is periodic of frequency ω_1. In that case the function H only depends on one variable x_1. But if the frequencies $\omega_1, \omega_2 \ldots \omega_n$ are not rationally dependent, the function h is not periodic and so a more general idea extending periodic is required. The easiest idea is the idea of quasiperiodic motion. This idea shares with periodic functions a Fourier series development. Indeed all smooth periodic functions $h(t)$ of period 2π can be written

$$h(t) = \sum_{n \in Z} a_n e^{int}$$

where a_n are the appropriate Fourier coefficients. While a general quasi-periodic function $h(t)$ frequency $\omega_1 \ldots \omega_n$ can be written

(1) $$h(t) = \sum_{|n| \geq 0} a_n e^{i\left(\sum_{i=1}^{N} \omega_i n_i\right)t}$$

where again a_n are appropriate generalized Fourier coefficients. See problems below.

Thus the question, already mentioned above, arises: What happens when such an integrable system is perturbed? This question was of course fundamental in the beginning of the twentieth century. Its partial resolution is due to Kolomogorov in the early 1950's who showed that for sufficiently small Hamiltonian perturbations satisfying certain irrationality and nondegeneracy conditions, a large number of quasiperiodic motions of integrable systems are preserved. In fact, a measure theoretic treatment of such preserved quasiperiodic motions can be undertaken. This work was later improved by Arnold and Moser who showed that the Kolomogorov analysis can be applied to many mechanical systems other than the usual ones, and that the perturbations involved need not be smooth.

In his book, <u>Sensations</u> <u>of</u> <u>Tone</u>, the great German scientist Helmholtz found the necessity to study quasiperiodic motions in attempting to explain "combination tones" in music. His explanation based on the nonlinearity of the ear has never been verified, partially due to mathematical reasons and partially because his understanding of the nonlinearity of the ear left many unresolved points. His idea was that combination tones arise as a response to the ear of an external driving force with two different frequencies, ω_1 and ω_2, with ω_1 and ω_2 not rationally related. Helmholtz's idea was that when the ear experienced these two external frequencies its response based on nonlinear Hamiltonian mechanics would involve a mixture of the frequencies ω_1 and ω_2, namely their sum and difference tones as well as higher harmonics. The summation tone would be of small amplitude and hence negligible. But on the other hand, the difference tone would be of substantial magnitude and would hence produce an audible effect on the eardrum.

Attempts to justify Helmholtz's ideas based on perturbation analysis have proved fruitless unless dissipation was introduced, because the series solution of the nonlinear differential equation so generated can be shown to diverge.

Here is a perturbation analysis of the Duffing equation with no dissipation and with a driving force $h(t)$ of two frequencies, say $h(t) = A \cos\omega_1 t + B \cos\omega_2 t$, showing how small divisors enter the problem. Consider the iteration

$$\ddot{x}_{n+1} + \alpha x_n + b x_n^3 = A\cos \omega_1 t + B\cos\omega_2 t$$

beginning with $x_0 = C \cos\omega_1 t + D \cos\omega_2 t$, where A, B, C and D denote constants. Then, at the first step, x_1 contains individual terms involving the cosine of $\omega_1, \omega_2, 3\omega_1, 3\omega_2$; but in addition, contains terms involving the cosine of $(\omega_1 \pm 2\omega_2)$ divided by squares of $(\omega_1 \pm 2\omega_2)$ itself. In the same way, the individual terms for x_{n+1} also involve terms containing higher and higher powers of

$$(\pm n\omega_1 \pm m\omega_2)^{-1} \quad \text{for n and m integers}.$$

Now if ω_1 / ω_2 is irrational, the general case, a famous result of Kronecker insures these latter terms come arbitrarily close to infinity for infinitely many integers m and n. From this fact we conclude that it is impossible to prove the convergence of such iteration procedures. These facts are known as the "problem of small divisors."

Thus the issue of great interest arises: In what sense do quasiperiodic motions for the relevant Hamiltonian mechanical systems actually exist? The so-called Kolomogorov-Arnold-Moser theory makes predictions for the Duffing equation mentioned above, but only subject to nonphysical restrictions and very small amplitude perturbations.

One approach to this issue taken up by both J. J. Stoker and K. Friedrichs was extending Hamiltonian systems by a small amount of added dissipation. This small amount of dissipation was then shown to generate convergence for the desired series expansions of solutions. The mathematical method involved is quite interesting and will be traced in the sequel. If the dissipation is sufficiently large, chaotic effects may take over and the resulting system no longer follows predictable behavior.

In this section, we attempt to make an additional contribution to the study of quasiperiodic motions by formulating new ideas. In the case of Helmholtz systems, to study such problems, our idea is to attempt to use the conservative nature of the system to get variational principles that can be utilized for the characterization of quasiperiodic motions. Another idea is to extend the concept of quasi-periodic motions to allow almost periodic oscillations with arbitrary frequencies, and then to allow the frequencies of the driving term to restrict these almost periodic motions, that satisfy the differential equation, to be quasiperiodic with the exact same frequencies of the driving term itself. We then attempt to utilize these extremal principles to demonstrate the existence of the desired motions both analytically and computationally. This approach will be described in a little more detail below.

Fourier Series Approach to Finding Quasiperiodic Motions of Finite Amplitude for Forced Nonlinear Dynamical Systems

A general quasiperiodic function $x(t)$ of frequencies $\omega_1 \ldots \omega_N$ can be written

(1)
$$x(t) = \sum_{|n| \geq 0} a_n e^{i\left(\sum_{i=1}^{N} \omega_i n_i\right)t}$$

where again a_n are the appropriate generalized Fourier coefficients, with the multi-index $n = (n_1, n_2, \ldots, n_N)$.

Based on this Fourier series development of a general quasi-periodic function $h(t)$, as mentioned above, we can attempt to find quasi-periodic motions of dynamical systems and thus to analyze the question (A). However this question is too general and it is necessary in order to utilize the Fourier series method to specialize our question somewhat. First it is important to ask how does one determine the frequencies $\omega_1, \omega_2 \ldots \omega_N$ of the unknown quasi-periodic motion? If one considers perturbation of a given quasi-periodic motion one can consider the frequencies of the perturbed motion as simply the original frequencies or small perturbations of them. In the Kologomorov approach, the frequencies of a small perturbation are held fixed. But for a general approach, it is necessary to specify the nonlinear dynamical system further. To this end, we consider forced dynamical systems with the force in term given by a function $h(t)$ that is quasiperiodic with given frequencies $\omega_1, \omega_2 \ldots \omega_N$ and we search for quasiperiodic motions of his dynamical system that have the same frequencies as the original forcing function. A typical example is the equation of Duffing with dissipation considered by Kurt Friedrichs. This equation can be written

(2)
$$\ddot{X} + c\dot{X} + X + \beta X^3 = h(t) \qquad [c \neq 0]$$

We suppose $h(t)$ is quasiperiodic and has the Fourier series development (1) given above with the frequencies $\omega_1, \omega_2 \ldots \omega_N$

independent over the rationals. Then we search for solutions of this Duffing equation $X(t)$ that have exactly the same frequencies but whose generalized Fourier coefficients a_n are uniquely determined by the parameters of the equation (2) itself.

Can one substitute the expression for $X(t)$ directly into the Duffing equation with given forcing function $h(t)$ and so determine the solution in terms of Fourier series expansions.? Is this solution a finite and well defined function and a solution of the Duffing equation? For given forcing function $h(t)$ of the form (1) is this solution unique? Can it be found explicitly? All these questions need to be answered if one hopes to understand quasi-periodic motions of dynamical systems. In order to understand this formulation of the problem of quasiperiodic motion we discuss below a well known formulation of Friedrich's and Stoker that uses simple functional analysis. The idea is to put quasi-periodic motions of the form (1) into a Banach space that we call X. It will be a space of quasiperiodic functions of fixed frequencies $\omega_1, \omega_2 \ldots \omega_N$. We define the norm in this space in terms of the Fourier components as

(3) $\| X(t) \|_X = \sum_n |a_n|$,   where n varies over all N-vectors

of integers Z^N and a_n is the relevant coefficient in the Fourier series expansion (1) above

This norm defines a Banach space and, in fact, it is easily shown by virtue of the equation (1) and the definition (3), that

$$\sup_t | X(t) | \leq \| X(t) \|_X$$

This inequality implies that convergence in the Banach space X implies uniform convergence. Now we consider the equation (2) in the Banach space X.

Consider the linear operator with $c \neq 0$,

$$Lx = \ddot{x} + c\dot{x} + x$$

and rewrite the equation (2) as

(4) $$Lx = h(t) - \beta x^3$$

Assume X and its first two derivatives are in the space X. Then we note by differentiating term by term that

$$\| Lx \| = \Sigma |r_n| |a_n|$$

where $r_n = 1 + icc_n - c_n^2$ with $c_n = \Sigma n_j \omega_j$.

It turns out that for c sufficiently small but not zero, the operator L is invertible in the space X with inverse L^{-1} satisfying the inequality

$$\| L^{-1}x \| \leq K \| x \|$$

For example, in case n = 2, we find for small c, namely $0 < c < \sqrt{2}$

$$K = c^{-1}(1 - \frac{c^2}{4})^{-\frac{1}{2}}$$

Consequently, equation (4) can be rewritten as

$$x = L^{-1}(h(t) - \beta x^3)$$

This equation can be solved by successive approximation, starting from $x_0 = 0$. The resulting sequence of quasiperiodic functions x_n converges in the Banach space X, provided the constants β and c, as well as the forcing term $h(t)$ are sufficiently small. This yields the desired quasiperiodic motion of the differential equation (1). However, it is important to notice that we cannot assume that the dissipation measured by the constant c vanishes, for then the operator L does not have an inverse.

Perturbations of Integrable Systems

The simplest perturbations of integrable Hamiltonian systems were thought to destroy all quasiperiodic motions until recently. Actually, the Russian mathematician Kolomogrov thought of a method for proving that most quasiperiodic motions for small perturbations preserving the Hamiltonian structure are not destroyed if the unperturbed system satisfies a nondegeneracy condition. Under these circumstances most quasi-periodic motions are not destroyed but are only slightly deformed. To make such systems precise, a large apparatus of technical results needs to be developed here. Since these ideas will not be needed in the sequel, we shall not develop this procedure.

However, it should be mentioned that this theory of preservation of quasi-periodic motions of integrable Hamiltonian systems under perturbation has many interesting applications in astronomy and in mechanics.

All these arguments are rather classical and require analogues of power series developments. The methods that we propose here are quite novel. In fact, these methods do not require any sort of local analysis, or any power series developments. Rather the arguments proceed globally from the form of the variational principles involved. They are directly connected with second order Hamiltonian systems, thus they do not discuss all the difficulties associated with dynamics, but only a small portion of them.

The most important mechanical problem to be considered here is the behavior of Hamiltonian systems near an equilibrium point. A Hamiltonian system that we envision can be written

(1) $\underline{\dot{w}} = J\underline{H}_w$ where $J = \begin{pmatrix} 0 & I \\ -I & 0 \end{pmatrix}$ and $\underline{H}_w$ denotes the gradient of H

Here $\underline{w}(t)$ is a vector with 2n components and J is the $2n \times 2n$ matrix, as indicated above.

We assume that the gradient of H vanishes at $w = 0$ which represents equilibrium for the mechanical system. Here $\underline{w}$ denotes the vector $\underline{w} = (w_1, \ldots, w_{2n})$. If this Hamiltonian system was integrable in the sense of Liouville described above, all motions of finite energy of the associated dynamical system would be quasi-periodic. Thus it is very interesting for the system (1) just described to see if this general non-integrable system also possesses quasi-periodic motions. An even simpler class of motions of the system (1) are the periodic ones.

Observation of nature leads one to study the phenomenon of periodicity. From a mathematical viewpoint it is necessary to analyze the notion of periodic functions and how they describe natural observations. It was Fourier in the nineteenth century who first thought of describing all periodic natural phenomena by Fourier series; that is, infinite series whose terms are composed of harmonics sin nt and cos nt. The work we are describing here on nonlinear science requires us to use Fourier's idea in a new mode. Indeed, we shall find function spaces composed entirely of periodic functions. When one desires to seek a given periodic motion in nonlinear science, one proceeds as follows. First one finds a governing equation or iteration scheme that describes exactly the natural system involved. Then one seeks a mapping between these function spaces whose solution is both periodic and also satisfies these governing equations at each point of space-time. Once this is accomplished, the full resources of mathematical analysis are at one's command to resolve the resulting scientific problem, usually a nonlinear operator equation and a function space. In the sequel we shall describe a few such instances and how our study differs from the traditional mode. Indeed, it is exactly here that our notion of periodic motion embodies both the ideas of Fourier and the ideas of modern analysis.

The simplest kind of motions for such a Hamiltonian system are periodic motions and here we use the fact that we are seeking periodic motions near an equilibrium so that we need only look at the associated linearized Hamiltonian system near $w = 0$. In fact we shall consider the matrix $JH''(0)$ and the associated linearized equation

$$(2) \qquad\qquad \dot{Z} = JH''(0)Z$$

It is important to consider the eigenvalues of this matrix and we shall assume that they are purely imaginary so that in fact the associated linearized differential equation has only bounded solutions. It is well known that in this case one can find a linear canonical coordinate transformation that reduces the quadratic part of the Hamiltonian H_2 to the form

$$(3) \qquad H_2 = \sum_{i-1}^{n} \frac{a_\nu}{2} (x_\nu^2 + y_\nu^2) \quad \text{where} \quad x_\nu = w_\nu, \quad y_\nu = w_{\nu+n}$$

$$\text{for } \nu = 1, 2 \ldots n$$

It is important to inquire whether the periodic solutions for this linearized system (2) give rise to periodic solutions for the full nonlinear system (1) near the equilibrium $w = 0$. If so, we can generate a large number of periodic solutions for the full nonlinear Hamiltonian system (1) that resemble normal modes of oscillation for ordinary mechanical systems. The basic result here is due to the Russian mathematician Liapunov and can be stated as follows.

Theorem (Liapunov) If a_i/a_j $(i = 1, 2, . . \hat{j} . . n)$ is not an integer, then the jth periodic motion of the linearized system (2) is a valid approximation to a periodic motion of the full nonlinear system (1). Moreover, the minimal period of the jth normal mode is near $2\pi/a_j$.

New methods, combining the calculus of variations and topology, have succeeded in greatly extending this result by removing the condition on the ratio of a_i/a_j altogether, provided the Hamiltonian function H has an appropriate convexity property near the equilibrium. This result will not be discussed further here, except to notice that the notion of Sobolev space of periodic functions and Ljusternik-Schnirelmann category are crucial ingredients for the proof. In order to discuss the first normal mode, however, topology is not needed, so we shall describe it in subsequent paragraphs.

This result on periodic motion describes only the normal modes, that is, the fundamental periodic solutions of the associated linearized system out of which all motions of the linear system can be constructed. Thus we are led to ask more generally, to find all solutions of the Hamiltonian system (1) near equilibrium. As mentioned earlier, Birkhoff considered this problem and attempted to make a coordinate transformation that would reduce the Hamiltonian system to the simpler Hamiltonian system described above. Research on this topic is still continuing.

A) Periodic Motions of Dynamical Systems of the Second Order

The periodicity of nature is a common occurrence. It is most important that mathematics attempt to reflect this periodicity by direct observation of the natural phenomena observed. It was Fourier (1768 – 1830), who described the well known representation of periodic motion by superposition of harmonics. This fundamental idea has had great utility throughout all mathematics and science to the present day.

Nonetheless, studies of periodic motions independent of Fourier's ideas have continued. The whole subject of celestial mechanics has as its centerpiece the study of periodic solutions of planets of the solar system and their perturbation by various astronomical effects. This has led to very precise charts of astronomical observations. The mathematical study of such periodic motions connects with the study of Hamiltonian dynamical systems. The methods for such studies relied on ingenious power series expansions, perturbation methods and fixed point theorems until twenty years ago, when I, and subsequently a number of others, tried to show the applicability of Fourier's ideas and the global calculation of variations arguments in work on celestial mechanics.

The basic idea is to study

first, the equations of motion for these systems of celestial mechanics in an abstract form; and

secondly, formulate specific problems about the motions of such systems. Recall here that one is not searching for the solutions of the traditional initial value problem, but rather, specific motions existing for all time;

thirdly, to formulate extremal problems to characterize such special motions;

fourthly, to utilize Fourier's series expansions of possible motions for the solution of the extremal problem. A case in point is the first normal mode for Hamiltonian systems going beyond Liapunov's theorem mentioned in the above section. Such a result would demand the complete removal of the condition $a_i/a_1 \neq$ an integer (i = 2, 3, ... n) mentioned there. Here the a_i are ordered by magnitude with a_1 being the largest. Such a result will be carried out for second order nonlinear Hamiltonian systems described below.

Problem   To find the analogue of the "first normal mode" for

$$\text{(4)} \qquad \frac{d^2x}{dt^2} + \nabla V(\underline{x}(t)) = 0$$

where $\underline{x}(t)$ is an N-vector function of the real variable t and V(x) is a strictly convex function of the vector variable $\underline{x}$.   The symbol ∇ denotes the vector operation, gradient, sometimes denoted grad.

We assume $\nabla V(x) = Ax + o(|x|)$ so that the linearized part of $\nabla V(x)$ at the equilibrium x = 0 is Ax. Here A is an N x N self-adjoint positive definite matrix.

In equation (4) we shall assume grad V(0) = 0, so that $\underline{x}(t) = 0$ is an equilibrium point for (4). Thus we are concerned in studying (4) with a great problem of classical mechanics: the study of solutions of (4) in the vicinity of an equilibrium point. In particular, we are concerned with periodic motions near this equilibrium.

The most direct mathematical analysis of equation (4) and the associated mechanical problems of periodicity reside in characterizing a periodic solution for this problem that in a certain sense has the smallest non-zero period. We shall formulate an extremal principle which has this type of periodic solution as its critical point. In order to determine this critical point it is natural to look at the Lagrangian formulation for this problem. See equation (6) below. Moreover, the mathematical analysis of equation (4) shows that the equation is "autonomous," i.e. does not depend explicitly on t. Consequently, the period of a solution of (4) is an additional unknown for the problem. Thus, we must insert the desired period for the periodic solutions as a parameter into (4).

The simplest analogue of (4), as mentioned above, is the associated linear system

$$\ddot{x} + Ax = 0$$

where A is an $N \times N$ self-adjoint positive definite matrix. The basic fact about this equation, as will be mentioned many times throughout this text, is that this system has N "normal modes," i.e. N linearly independent periodic solutions. These solutions form a natural basis for all solutions of the equation. Thus they can be considered fundamental building blocks. Consequently, it is very important to find their "nonlinear analogues." When one studies the calculus of variations formulation for (4), one observes that the periodic solution corresponding to the first normal mode is not an absolute minimum of the Lagrangian. (See problems below). Thus it is necessary to find some mathematical approach to study saddle points of the associated Lagrangian. The easiest way to do this is the method of "natural constraints" studied in this book. To this end, we proceed as follows utilizing the fact that the equation is autonomous.

To find the natural constraint for this problem we make a change of variables by setting $t = \lambda s$ in (4) and seek 2π periodic solutions in s. Such solutions correspond to $2\pi\lambda$ periodic solutions in t. Thus the parameter λ is a parameter measuring period.

The transformed equation can be written in terms of s as follows:

(4')
$$\frac{d^2x}{ds^2} + \lambda^2 \nabla V(x) = 0$$

Integrating the transformed equation over a period we find that a constraint for the problem is

(5)
$$\int_0^{2\pi} \nabla V(\underline{x}(s))ds = 0$$

Here we have used the simple fact that the integral of the first term over a period vanishes because of the periodic nature of the desired solutions, i.e. $\underline{x}(s)$ and all its derivatives satisfy 2π periodic boundary conditions; so for example, $\underline{x}(0) = \underline{x}(2\pi)$ with the same equation holding for all derivatives of $\underline{x}(s)$.

The natural Lagrangian for the transformed problem is

(6)
$$L(\underline{x}(s)) = \frac{1}{2}\int_0^{2\pi} \left|\underline{\dot{x}}(s)\right|^2 - \lambda^2\int_0^{2\pi} v(\underline{x}(s))$$

and is unbounded above and below (see problems) when studied over the class of 2π periodic vector functions $\underline{x}(s)$ so that all maxima or minima of L are infinite. However, the method of natural constraints suggests that one uses the constraint defined by equation (5) in attempting to find critical points of (6). This simply means that one regards the parameter λ^2 as a Lagrange multiplier to be determined a posteriori, and associates with (6) a constrained variational problem as follows.

Consider the isoperimetric variational principle of minimizing the "kinetic energy" over the constraint Σ.

$$(7) \qquad K(\underline{x}(s)) = \int_0^{2\pi} \left| \frac{d\underline{x}}{ds} \right|^2$$

$$\text{where } \Sigma = \left\{ \int_0^{2\pi} V(\underline{x}(s))ds = R, \int_0^{2\pi} \nabla V(\underline{x}(s))ds = 0 \right\}$$

In this variational principle the period λ appears as a Lagrange multiplier and the solutions of the isoperimetric variational principle automatically yield a nonconstant periodic solution (indeed, the desired minimizing vector function has nonzero kinetic energy because $R \neq 0$). Here the vector functions $x(t)$ in Σ are defined relative to the Sobolev space $W_{1,2}((0, 2\pi), \mathcal{R}^N)$, a space consisting of N-vector 2π periodic functions $\underline{x}(s)$ with each component $x_i(s)$ an element of the appropriate Sobolev space of periodic functions $W_{1,2}(0, 2\pi)$.

What can be said about the solution of this isoperimetric variational problem? First, the minimum of the problem exists and is attained at a periodic solution $x_R(t)$. This periodic solution is smooth, and as R varies over all real numbers, one parameter families of periodic solutions are obtained. This family of solutions is computable by the methods of constrained optimization. Moreover, the parameter λ that occurs can be shown to be the minimal period of the periodic motion. Finally, assume $\nabla V(\underline{x}) = \sum_i \lambda_i \underline{x}_{i-i} + o(|\underline{x}|^2)$. Then as R $\to$ 0 the solution and its minimal period $(\underline{x}_R(S), \lambda) \to (0, 2\pi/\lambda_N)$. This last fact signifies that the periodic solution of the isoperimetric variational problem (7) is the analogue of the first normal mode that we are seeking since its period is close to the period of the first normal mode of the associated linearized system at $\underline{x} = 0$. The important thing to notice in this case is that, contrary to all power series developments, there is no small divisor restriction on the existence of the one parameter family of periodic solutions emanating from the first normal mode of the linearized problem. This calculus of variations ideas, an extension of the famous Liapunov's theorem for periodic solutions near an equilibrium point, as was mentioned in the previous section.

Significance of the Results Found

Let us consider the importance of these results as stated briefly above. The results show that there is a global constructive method for determining the analogue of the first normal mode of certain linear Hamiltonian systems. This first mode is a direct extension of the problem associated with the linear Hamiltonian system

$$\frac{d^2 \underline{x}}{dt^2} + A\underline{x} = 0$$

It shows also that the method of determining this nonlinear normal mode is a global calculus of variations problem since the conventional approaches are perturbative and utilize power series expansion techniques. Moreover, the idea does not require that we are near the stationary point $\underline{x} = 0$, but rather requires only a global geometric property of the Hamiltonian potential $V(\underline{x})$, namely the strict convexity of the function $V(\underline{x})$.

Here are some proofs of some of the important results that I have stated above for this calculus of variations problem:

1)   The   constraint   given   by   (5) $\displaystyle\int_0^{2\pi} \nabla V(\underline{x}(s))ds = 0$

is a natural one for the first normal mode of (4) in case the potential $V(x)$ is strictly convex. This means adding the constraint (5) to the isoperimetric problem ( ) does not affect solutions of this variational problem.

Proof   The key idea here is showing that the smooth critical points $\underline{x}(s)$ of the Lagrangian (6) are unaffected by the constraint (5). Notice that the constraint (5) is a system of N nonlinear equations associated with N Lagrange multipliers $\beta = (\beta_1, \beta_2, \ldots \beta_N)$. The associated Euler-Lagrange equations for the extremals of (7) are

(7*) $x'' + \lambda_i \nabla V + \beta \cdot V(\nabla V) = 0$

Thus integrating this last equation over a period $(0, 2\pi)$ we find the N equations

$$\beta \cdot \int_0^{2\pi} \nabla(\nabla V) = 0$$

The strict convexity of V then forces the constant vector $\beta = 0$.

Moreover, the strict convexity of V also forces the Lagrange multiplier λ_1 to be positive so that $\lambda_i = \lambda^2$ for some real nonzero number λ. To see this, set $\beta = 0$ in (7*) and take the inner product with $x(s)$ and then integrate over the period $(0, 2\pi)$ to find

$$\int_0^{2\pi} |x|^2 ds = \lambda_1 \int_0^{2\pi} \nabla V(\underline{x}) \cdot \underline{x}(s) ds$$

2) The variational problem (7) has a smooth periodic solution when it is confined to the Sobolev space $W_{1,2}((0, 2\pi), \mathbf{R}^N)$.

Proof   This result is attained by analyzing the isoperimetric problem (7) is a combination of Sobolev space ideas and the utilization of Hilbert space techniques of weak convergence. The key idea is that the natural constraint (5) uniquely determines the constant vector part of any admissible vector function $x(s)$ in the constraint set Σ in terms of its component with mean value zero. In other words, N directions, causing a saddle point to appear are excised by the natural constraint. Consequently standard methods of the direct methods of the calculus of variations (see my book [3], chapter 6) lead directly to the attainment of a smooth extremal. In short, the argument goes as follows: One simply takes a minimizing sequence $x_n(s)$, $n = 1, 2, 3 \ldots$ for the problem (7) and proves by the relevant compactness theorems of Sobolev that the minimizing sequence can be refined to a subsequence that converges weakly to the unique element $x_\infty(s)$ in the Sobolev space mentioned above. This element lies on the constraint set Σ by the weak continuity of the functionals defining the set Σ in the Sobolev space and also is the desired extremum. Standard regularity methods show that $x_\infty(s)$ can be assumed to be smooth so that $x_\infty(s)$ is the desired smooth periodic solution.

The case of convex potentials (not necessarily strictly convex) is handled by an approximation argument.

3) The Lagrange multiplier λ that occurs in the variational problem (7) is directly related to the minimal period of the periodic solution found for (4).

Proof This result follows by supposing the extremal $x(t)$ does not have minimal period $2\pi\lambda$. Then in terms of the independent variable s, for integer n, $x(s/n) \in \Sigma$ would be a periodic extremal of the constrained variational problem (7). However, this contradicts the minimality of the $x_R(s)$ as a global minimum (7) with the constraint Σ.

4) Assuming $V(x) = \frac{1}{2}\Sigma\lambda_i^2 x_i^2 + 0\,(|x|^3)$, the periodic solution $x_R(s)$ and the minimal period $2\pi\lambda$ have the following properties,

$(x_R(s), \ 2\pi\lambda) \rightarrow (0, \ 2\pi/\lambda_N)$ where λ_N is the largest of the eigenvalues λ_i i.e. the minimal period of $x_R(s)$ tends to the minimal period of the "first" normal mode of the linearized problem

$$\frac{d^2x}{dt^2} + V''(0)x = 0$$

Proof   To prove this result we merely let $R \to 0$ in (7)   and notice that the quadratic terms in   (7)   are then dominant.   In particular, as $R \to 0$ $|x_R(s)| \to 0$ and the lowest   eigenvalue $2\pi/\lambda_N$)   of the quadratic variational problem

$$\inf \int_0^{2\pi} |\dot{x}(s)|^2 \text{ over } \tilde{\Sigma} = \left\{ \int_0^{2\pi} \sum_{i=1}^{N} \lambda_1 x_1^2 = 1, \int_0^{2\pi} x = 0 \right\}$$

is the correct approximation for the minimal period $\lambda(R)$ as $R \to$ 0, in (7).

B) Quasiperiodic Motions of Second Order Systems

Introduction

As stated earlier in this chapter, the systems that are normally considered the simplest are called integrable systems. These integrable systems for nonlinear dynamics are always Hamiltonian. As stated above, their integrability is based on a theorem of Liouville which states that for Hamiltonian systems with N degrees of freedom a system is integrable if it has N conservation laws given by the equation $F_i = 0$, $i = 1,2,3 \ldots N$. Moreover, these conservation laws are required to be independent and such that these are in involution., i.e. the Poisson bracket of any two of these conservation laws vanish.

Now we wish to address the question of the perturbation of integrable systems. Our first observation is that such integrable systems, defined by Liouville's theorem, always possess quasi-periodic solutions. In fact, we can say that all finite energy solutions of these integrable systems are quasiperiodic. A good example is the family of quasiperiodic solutions generated by N coupled harmonic oscillators. This system can be written in the form

$$(1) \qquad \underline{\ddot{x}} + A\underline{x} = 0$$

Here $x(t)$ is an N vector of functions, and A is a $N \times N$ matrix which is non-singular positive definite and self-adjoint. Periodic motions for such systems were called normal modes in the sections directly above, but general solutions are finite linear combinations of these normal modes. Hence, in general, any solution of this system (1), will be quasiperiodic.

The perturbation question that I wish to address can be stated as:

Are quasiperiodic solutions of the perturbed integrable system preserved or completely destroyed?

This question cannot be tackled by classical methods. It is both local and global in nature, since it is not clear that small amplitude solutions only, are considered. Before we proceed further it is best to display the definition of quasiperiodic functions.

Definition   A continuous function $q(t)$ is called quasiperiodic with frequencies $\omega_1, \omega_2, \ldots \omega_N$ if there is a function continuous Q defined on the $R^N \rightarrow R^1$, denoted $Q(x_1, x_2, \ldots x_N)$ such that the continuous function Q is 2π periodic in each coordinate x_i and moreover

(2) $q(t) = Q(\omega_1 t, \omega_2 t, \ldots \omega_N t)$.

I refer to the function $Q(x_1, x_2, \ldots x_N)$ as the generating function for the quasi-periodic function $q(t)$. It will be important in the sequel to realize that the determination of this function Q can be made the abstract basis for finding the quasiperiodic function $q(t)$.

This definition is important because it shows that to find a quasi-periodic function it is necessary to find functions on R^N that are periodic in each coordinate. Another way of saying this is that one considers a torus T^N in N dimensions and seeks for real-valued function on T^N.

A New Optimization Procedure for Quasiperiodic Motions of Nonlinear Dynamical Systems

We begin as in the last section with the study of the Duffing equation. In the case of quasiperiodic forcing and no dissipation, this equation can be written

$$\ddot{q} + aq + bq^3 = h(t) \quad \text{where} \quad \ddot{} = d^2/dt^2$$

In order to analyze this question it is important to have a more careful definition of quasiperiodic. As mentioned above, this idea of quasiperiodic can be better defined in terms of a function of N variables. We say here a function is quasiperiodic if there is a function of N variables $H(t_1, t_2, t_3 \ldots t_N)$ called the generating function for the quasiperiodic $h(t)$ that has certain periodicity properties.

The periodicity properties are these. Suppose we want to define a quasiperiodic function of N frequencies $\omega_1, \omega_2 \ldots \omega_N$. Then we say that the function $h(t)$ is quasiperiodic if there is a generating function $H(t_1, t_2, t_3 \ldots t_N)$ with the property that

$$h(t) = H(\omega_1 t, \omega_2 t, \omega_3 t \ldots \omega_N t)$$

and the function $H(t_1, t_2, t_3 \ldots t_N)$ is 2π periodic in each variable. Thus we connect the quasiperiodic function with a generating function of N variables that is 2π periodic in each variable. In this sense the quasiperiodic function $h(t)$ is connected with the generating function H defined on the torus T^N Consequently the search for quasiperiodic solutions $x(t)$ in the same way can be reduced to a question of finding the generating function for $x(t)$, namely $X(t_1, t_2, t_3 \ldots t_N)$ defined on T^N.

A case in point is solving the Duffing equation with quasi-periodic right hand side h(t) depending on N frequencies $\omega_1 . . . \omega_N$. In order to solve this equation we need only determine the generating function for the solution in question.   Once this generating function has been determined we automatically find the solution itself via the prescription given above.

If the original differential equation has no dissipation, such as our Duffing equation, the associated problem for the generating function will involve partial differential equations that reflect no dissipation.   This signifies that the solution of the partial differential equation called PDE can be obtained by extremal principles for the Duffing equation we have in mind.   In certain cases, the solution will be obtained as an absolute minimum and consequently can be found numerically by optimization procedures.

At this stage we digress to consider the current situation in nonlinear dynamics.   Generally one considers what happens when nonlinear systems are studied with a given initial condition.   If these systems are defined by an iteration process and depend on a parameter, say

(3) $x_{n+1} = \lambda F(x_n)$

Then it is known for a large class of function   F   even in one dimension that when the parameter λ gets larger the system exhibits chaotic dynamics (see Problems);   this means that such systems exhibit sensitivity to perturbations to initial data and are unpredictable asymptotically.   Such chaotic systems may exhibit bifurcation phenomenon in addition to chaotic dynamics, but in any case it is important to note that the initial value problem is not the object of study in this chapter on nonintegrability.   Indeed when one passes to systems of ordinary differential equations one would like to study the initial value problem and predict the outcome of solutions after a long time. For this purpose the notion of an integrable system has been very helpful, since for such systems explicit solutions based on integrability by quadrature are available.

Now we inquire what happens when such systems are perturbed. The first point to note is that we assume throughout that the perturbation considered preserves the Hamiltonian structure. The KAM theorem, discussed briefly below, is applicable provided the perturbation is sufficiently small and a certain degeneracy is excluded. Otherwise there is virtually nothing known about such perturbations.

Loosely speaking, the KAM theorem starts with certain nondegenerate integrable Hamiltonian systems with N degrees of freedom that can be written in the form

$$\dot{p}_i = -\frac{\partial H}{\partial q_i} \quad , \quad \dot{q}_i = \frac{\partial H}{\partial p_i} \qquad i = 1, 2 \ldots N$$

Here $H(p, q)$ is a real-valued function of the 2N variables with $p = (p_1, p_2 \ldots, p_N)$ and $q = (q_1, q_2, \ldots, q_N)$. One then considers a small Hamiltonian perturbation for such systems, so that the Hamiltonian H is changed to $H + \varepsilon H_1$. Then the result states that if one excludes certain resonant cases and requires the perturbation to be sufficiently smooth and sufficiently small, many of the quasi-periodic solutions of the unperturbed Hamiltonian system remain quasi-periodic after the perturbation.

In this section we propose a new approach to studying quasi-periodic motion of certain Hamiltonian dynamical systems based on optimization techniques. The technique that I propose is new, yet because of the advances in optimization, promises computability as well as insight into the nonlinear process involved. Research discussed here represents joint work with my colleague Alexander Eydeland.

A Simple New Example to Find Quasiperiodic Motions

Our idea is to write down a standard nonlinear differential second order equation that may have quasiperiodic solutions. We then find from this nonlinear ordinary differential equation the associated nonlinear partial differential equation for the generating function (as described above) whose solution will yield the function of n variables that we seek. Because the system is Hamiltonian we are then able to find a variational principle for this function of n variables. As a final step we show that in certain cases this function of n variables can be determined by infinite dimensional minimization techniques. Here is a simple case in point.

Let a and b be positive constants. We consider the nonlinear ordinary differential equation (a special case of Duffing's equation)

$$(1) \qquad\qquad \ddot{q} - aq - bq^3 = h(t)$$

where $h(t)$ is a given quasiperiodic function of frequencies $(\omega_1, \omega_2, \ldots \omega_n)$. We seek a variational principle that determines a quasi-periodic solution $q(t)$ of frequencies $\omega_1, \omega_2, \ldots \omega_n$ for (1) assuming $h(t)$ is given as above. (i.e. we seek a quasi-periodic solution $q(t)$ whose frequencies are exactly identical with those of the forcing term $h(t)$. In symbols, this means we seek a generating function $u(t_1, t_2, \ldots t_n)$ of n variables 2π periodic in each t_i (i = 1, 2, ... n) with

(2)
$$q(t) = u(\omega_1 t, \omega_2 t, \ldots \omega_n t)$$

We now derive a relationship between the function $q(t)$ and $u(t_1, t_2, \ldots t_n)$. Differentiating (2) with respect to t we find

(3)
$$\dot{q}(t) = \sum_{i=1}^{n} \omega_i u_i(\omega_1 t, \omega_2 t, \ldots \omega_n t)$$

Differentiating once again we find

(4)
$$\ddot{q}(t) = \sum_{i,j=1}^{n} \omega_i \omega_j u_{t_i t_j}$$

$$= Lu$$

Here Lu denotes the second order differential operator defined by the right hand side of (4). This operator L is self-adjoint and positive semi-definite when regarded as a second order differential operator on $\mathfrak{R}^n$. Indeed, the characteristic form of L, $L_0(k)$ can be written in terms of the vector $k = (k_1, k_2, \ldots k_n)$ as

$$L_0(k) = \left(\sum_{i=1}^{n} \omega_i k_i \right)^2 \geq 0$$

This form is semidefinite (for $n>1$), since $L_0(k)$, can be written in terms of the vector $k = (k_1, k_2 \ldots k_n)$ as

$$L_0(k) = \left(\sum_{i=1}^{n} \omega_i k_i \right)^2 \geq 0$$

This form is semidefinite (for $n > 1$), since $L_0(\xi) = 0$ whenever

$$\sum_{i=1}^{n} \omega_i \xi_i = 0 \quad \text{(i.e. on a hyperplane of codimension 1).}$$

We also note that the operator L is the square of a first order differential operator $M(u)$ defined as follows

$$Mu = \sum_{i=1}^{n} \omega_i u_{x_i}$$

Consider then the following partial differential equation associated with (1).

(5) $$Lu - au - bu^3 = H(t_1, t_2 \ldots t_n)$$

Consider now the functional on the Sobolev space H of odd functions in $W_{1,2}([-\Pi,\Pi])$

(6) $$\Phi(u) = \int_{-\Pi}^{\Pi}\ldots\int_{-\Pi}^{\Pi}\left[Mu.Mu + au^2 + \frac{1}{2}bu^4 + 2Hu\right]dt_1\,dt_2\ldots dt_n$$

Lemma 1   If $u(x_1,x_2 \ldots x_n)$ is a smooth critical point of the functional $\Phi(u)$ on H, then the function

$$q(t) = u(\omega_1 t,\omega_2 t,\ldots \omega_n t)$$

is a solution of the problem (1).

Proof   Indeed $u_1(t_1, t_2 \ldots t_n)$ must satisfy the Euler Lagrange equation for the functional $\Phi(u)$. Thus $q(t)$, defined by (2), must satisfy (1).

Higher Dimensional Cases

This method just discussed of course is quite general; it can be applied to higher dimensional cases equally well. However, for simplicity we shall limit our discussion here to second order Hamiltonian systems for forcing term so that the frequencies of the known oscillators discussed will be clearly defined. Such systems occur naturally in many mechanical situations, but their numerical computations have not yet been carried out. The system we discuss here can be written as the quasiperiodically forced Hamiltonian system

$$(7) \qquad\qquad \ddot{\underline{x}} - \nabla U(\underline{x}) = \underline{h}^{(N)}(t)$$

where $\underline{x}(t)$ is a   N vector function of t, and $\underline{h}^{(N)}(t)$ is an N vector of quasiperiodic functions with components each of the form (1) and $U(x)$ is an even function x with $U(x) / |x| \rightarrow \infty$ as $|x| \rightarrow \infty$. Associated with (7) we can consider a system of nonlinear partial differential equations extending the equation (4). The solution of the resulting system can be characterized by utilizing optimization processes and the calculus of variations, but we shall leave this topic to another occasion.

Free Quasiperiodic Vibrations

We discuss here a beginning idea that will be elaborated in later sections. Consider quasiperiodic motion for linear systems without dissipation and without forcing term consisting of the motion of n coupled harmonic oscillators. Here the equation of motion for the couple system can be written

(1) $\qquad \ddot{x} + Ax = 0$ where A is a self-adjoint positive definite matrix with eigenvalues $\lambda_1^2, \lambda_2^2, \lambda_3^2 \ldots \lambda_n^2$

It is well known that this system has n distinct normal modes of period $2\pi/\lambda_1, 2\pi/\lambda_2, \ldots 2\pi/\lambda_n$. Moreover, every motion of the coupled system consists of a finite linear combination of these normal modes. This gives rise to the fact that a general vibration of a system is a quasi-periodic motion with N frequencies $2\pi/\lambda_1, 2\pi/\lambda_2, \ldots 2\pi/\lambda_N$.

Now we inquire about the quasiperiodic motions of a nonlinear conservative system without forcing term. Here we merely derive a new variational principle that may yield a new approach to studying such quasi-periodic motions for nonlinear free vibrations. Such a given system can be represented by the equation

(2) $\qquad \underline{\ddot{x}} + A\underline{x} + N\underline{x} = 0$ where $N\underline{x}$ represents higher order terms in the displacements $x_1, x_2 \ldots x_N$

We assume that the system is conservative. That means that it suffers no loss of energy due to dissipation. The significance of this fact is that the nonlinear term $N(x)$ is the gradient of a scalar function $U(x)$ and we write

(3) $\qquad\qquad N\underline{x} = \text{grad } U(x) = \left(\dfrac{\partial U}{\partial x_1}, \dfrac{\partial U}{\partial x_2}, \dfrac{\partial U}{\partial x_n} \right)$

Here the function $U(x)$ is known as a potential function since one imagines that it represents the potential energy due to nonlinear displacements of a mechanical system with n degrees of freedom. In order to analyze this quasiperiodic solution of equation (2) we transform the ordinary differential equation as given into a partial differential equation on the n-dimensional torus, T^n. This equation represents a partial differential equation with boundary conditions for the generating function $X(x_1, x_2 \ldots x_n)$. It can be written in the form

(4) $\qquad L\underline{X} + A\underline{X} + N(\underline{X}) = 0 \qquad$ where L is the linear differential operator given by the formula:

(5) $\qquad L\underline{X} = \displaystyle\sum_{i,j=1}^{n} \omega_i \omega_j \underline{X}_{x_i x_j}$

Here the vectors $L\underline{X}$ have n components and the partial differential equation (4) is supplemented by periodic boundary conditions at the ends on the interval $(-\pi, \pi)$. Moreover, to introduce a time scale, fix the frequencies of the $\underline{X}(t)$ by introducing the parameter λ defined by

$$t_i = \lambda s_i \qquad (i = 1, \ldots n)$$

Note by our assumptions, that (4) can be rewritten

(4') $\qquad\qquad L\underline{X} + \lambda^2 \nabla V(\underline{X}) = 0$

where $\qquad\qquad \nabla V(\underline{X}) = A\underline{X} + \nabla U(\underline{X}).$

Moreover, for the sequel, we assume that the function $V(\underline{X})$ is strictly convex.

Then by integrating (4') over T^N and using the 2π periodic boundary conditions on each side of T^N, we find for any smooth solution $\underline{X}(s)$ of (4')

(6)
$$\int_{T_N} \nabla V \underline{X}(s) ds_1 ds_2 \ldots ds_n = 0$$

To analyze the system (4'), we observe with the associated boundary conditions the operator L is self-adjoint. Thus we can find a variational principle for the solutions of the equation (4). This principle can be used as an effective method to find the solutions of (4) both analytically and numerically. The principle can be written as follows:

Minimize the quadratic form

$$(L\underline{X}, \underline{X})_{L_2} = \sum_{i,j} \int_{-\pi}^{\pi} \omega_i \, \omega_j \, \dot{\underline{X}}_{x_i} \cdot \dot{\underline{X}}_{x_j}$$

subject to the constraint defined by (6) and the following equation defined for R positive:

(7)
$$R = \int_{-\pi}^{\pi} \int_{-\pi}^{\pi} \cdots \int_{-\pi}^{\pi} \{ \frac{1}{2} A \underline{X} \cdot \underline{X} + U(\underline{X}) \} ds_1 ds_2 \cdots ds_n$$

This isoperimetric characterization leads to a global family of periodic solutions for $n = 1$, as will be shown in the sequel. For $n > 1$, it is still undetermined whether quasiperiodic free vibrations of (2) will be obtained by this approach.

Section 1.8   Almost Periodic Motions of Dynamical Systems

In this section we wish to extend the notion of quasiperiodic motion described above by utilizing a theory first developed by Harald Bohr. We begin with a purely mathematical problem as first stated by Bohr in his famous book on almost periodic functions, "Which functions $f(x)$ can be resolved into pure vibrations on the real axis R^1 (i.e. $-\infty < x < \infty$), i.e. are "representable" by a trigonometric series of the form $\Sigma A_n \exp(i\lambda_n x)$?" Here the numbers λ_n are arbitrary real numbers as contrasted with Fourier series in which the numbers λ_n are integers. This infinite series just described is referred to as a Fourier series for an almost periodic function. Bohr determined exactly which continuous functions defined on the real line have Fourier series of the above form. He called these "almost periodic functions." Bohr's fundamental theorem was a statement describing exactly which continuous functions on the real line have Fourier series in the above sense. Of course, the basic question is one of convergence, and Bohr was also able to solve this problem relative to continuous functions and uniform convergence.

The simplest way to define almost periodic functions is to first define trigonometric polynomials.

Definition   A trigonometric polynomial on the real line is simply a finite sum of the form

$$f(x) = \sum_{1}^{n} a_j \exp(i\lambda_j x), \quad \text{where } \lambda_j \text{ are real numbers called frequencies}$$

Almost periodic functions in the sense of Bohr here called {U.A.P.} are simply the uniform limits of trigonometric polynomials.

In particular, we wish to consider how one obtains almost periodic solutions of nonlinear Hamiltonian systems of differential equations using the abstract techniques described in this book. Indeed, a fundamental problem in dynamical systems is to replace the study of the initial value problem with the search for special solutions of a given nonlinear dynamical problem. In this way, one hopes to avoid chaotic dynamics and search for a well defined mathematical problem that has a well defined solution capable of study both by analytical and numerical means.

For example, is it possible to use the calculus of variations techniques described above for periodic and quasiperiodic motions of Hamiltonian systems when we do not specify the desired frequencies λ_n in advance? To this end one hopes to use the beautifully developed theory on almost periodic functions described by Bohr. However, there is a problem in this connection because Bohr always dealt with functions that were continuous on the real line and generally considered arguments involving uniform convergence. Here is a case in point that is a direct analogue of the quasi-periodic problem considered in the last section.

Suppose one is given an almost periodic forcing term $h(t)$ satisfying the Duffing equation

(*)
$$\ddot{x} - ax - bx^3 = h(t)$$

where a and b are positive constants.

Question: Does this equation have an almost periodic solution with the same Fourier series frequencies as $h(t)$?

Very relevant points in answering this question are:

a) The equation (*) possesses a uniformly almost periodic solution $x(t)$.

b) The uniformly almost periodic solution of equation (*) is unique in the class of uniformly almost periodic functions. (See Appendix 4 for the appropriate definitions).

c) Any uniformly almost periodic solution $x(t)$ of equation (*) satisfies the following estimate:

$$\sup_{t \in \mathscr{R}^1} |x(t)| \leq C \sup_{t \in \mathscr{R}^1} |h(t)|$$

where C is an absolute constant.

These results allow the Fourier exponents of the solution $x(t)$ to be connected with the Fourier exponents of the forcing term $h(t)$ in a very direct manner. In particular, if $h(t)$ is quasiperiodic with N exponents independent over the rationals, one can deduce that the unique solution $x(t)$ is also quasiperiodic with the same Fourier exponents as $h(t)$.

A basic question in utilizing the abstract set-up of this book is the practical utilization of calculus of variations techniques. To this end it is necessary to find appropriate Hilbert space structures of almost periodic functions extending the original spaces considered by Bohr; minimization arguments, for example, generally require such completions. In this way, we fit advances in mathematical analysis and nonlinear science into the almost periodic context. In the Appendix 4 included at the end of this chapter, one will find special ways of defining Hilbert space structures on the space of almost periodic functions when properly extended. We are certain that such Hilbert space structures will be crucial for finding large amplitude almost periodic motions of nonlinear dynamical systems.

It is also possible to analyze quasiperiodic motions with large amplitude without specifying their frequencies by passing to the larger class of almost periodic functions. This idea is novel, but nonetheless, it has been recently utilized by myself in joint work with Y. Y. Chen with some success. The idea relative to the forced Duffing equation, is simply to look for a solution in the class of almost periodic functions. Assume this can be achieved in the class of continuous functions (u.a.p.). By using the fundamental theorem of Harald Bohr, one knows that the solution can be represented as a generalized Fourier series with an infinite number of frequencies of the form given above. Indeed, one can make a careful study of finding the relationship between the frequencies of the function $h(t)$ and the solution $x(t)$ for the forced Duffing equation under consideration, assuming the solution $x(t)$ is unique. In fact, one finds that the frequencies of $h(t)$ are reflected in the solution $x(t)$ in that their Fourier exponents are closely related.

This fact has an important consequence for quasiperiodic motions of the forced Duffing equation

$$d^2x/dt^2 - ax - bx^3 = h(t)$$

where a and b are positive constants.

It implies first that if $h(t)$ is quasiperiodic, with frequencies ω_1, $\omega_2, \ldots \omega_n$ (with these frequencies independent over the rationals) then the solution $x(t)$ is also quasiperiodic with exactly the same frequencies $\omega_1, \omega_2, \ldots \omega_n$. Secondly, it implies, since the solution of the forced Duffing equation is unique, that minimization techniques can be used to find and compute this quasiperiodic motion. Indeed, consider the functional

$$I(x) = \lim_{T \to \infty} \frac{1}{2T} \int_{-T}^{T} \left[\dot{x}^2 + x^2 + \frac{x^4}{2} + 2h(t)x \right] dt$$

This functional can be considered defined over the class of almost periodic functions $x(t)$ in the Hilbert space H with the Besicovitch-Sobolev norm

$$\| H \|_H^2 = \lim_{T \to \infty} \frac{1}{2T} \int_{-T}^{T} (\dot{x}^2 + x^2) dt$$

This functional is strictly convex in the Hilbert space H as well as lower semi-continuous with respect to weak convergence in H and moreover, is "coercive" in the sense that $I(x) \to \infty$ as norm $\|x\|_H \to \infty$. Consequently, this functional has a unique absolute minimum $\bar{x}$ in H. This minimum has a generalized Fourier series development that coincides with the continuous almost periodic solution solution of the forced Duffing equation just obtained . Consequently, this minimum will be smooth and the desired solution of the forced Duffing equation. The full details of this point will be published in a separate paper.

Appendix 1   Nonlinear Fredholm Operators

What are the abstract properties essential to carry out the different types of nonlinear processes discussed in this book? Certainly the classes of nonlinear equations to be discussed are varied and of differing types. Thus to find unity in this diversity, it is necessary to reason on an abstract basis.

We carry out such arguments as follows. We consider a mapping A between two function spaces denoted X and Y. We shall assume throughout this book that these function spaces are Banach spaces i.e. complete normed linear spaces whose metric is given by the formula

$$d(x, y) = \|x - y\|$$

The mapping between these spaces is assumed to be differentiable at a point x and we compute its linearization, i.e. its Frechet derivative at the point x which we denote by A'(x). (See the material immediately following in the next Chapter, Section 2.1).

Now, we ask the question, what type of operator A   will possess the abstract properties needed to carry out the arguments of the different processes discussed in this book. At the moment, it is believed the way to proceed with this question is to look at the linearization A'(x), assumed to depend smoothly on x. This reduces the question to the analysis of linear operators. Since this topic has been very well studied, it makes good sense to utilize the well-known linear results concerning the classification of operators at this stage. And in fact, we consider an abstract operator L acting between the same two spaces as above, X and Y. Then the fundamental question of linear analysis concerns solving the nonhomogeneous operator equation

$$Lx = y$$

when y is a fixed element of the Banach space Y. The solvability of this equation is most easily discussed by assuming that the operator L is Fredholm, i.e. that the well-known Fredholm alternative applies to it. Thus we make the following

Definition   A bounded linear operator L acting between the Banach spaces X and Y is Fredholm if

    i) L has closed range
    ii) dim ker L is finite dimensional
    iii) dim coker L is finite dimensional

A C^1 nonlinear operator A acting between the Banach spaces X and Y is Fredholm if the linear operator A'(x) is a Fredholm linear operator for each x an element of X.

Definition   The index of a bounded linear Fredholm operator L acting between the two spaces X and Y is defined to be the integer

$$index\ of\ L = dim\ ker\ L - dim\ coker\ L$$

The index of a C^1 nonlinear operator A acting between the spaces X and Y is defined to be the index of the linear operator index A'(x) for any x in X. The key fact here is that the index of A does not depend upon x as x varies over the space X. The reason for this fact is that the operator A depends in a continuous manner on x, thus, since the index of A is always an integer, and the index must depend continuously on A', the index of A must be a continuous function of x, and so an integer, and this integer must coincide with index $A'(x)$.

There are various inequalities that guarantee that a bounded operator L acting between two spaces, X and Y, is Fredholm. These inequalities can be written as follows

(1) $$\|y\| \leq c_1 \|f'(x)y\| + |y|_0 \ ,$$

(2) $$\|y\| \leq c_2 \|f'*(x)y\| + |y|_1 ,$$

where the constants c_1 and c_2 are independent of y and $|y|_0$ and $|y|_1$ are compact seminorms defined on Y.

The notion of a Fredholm operator is very useful in studying the singular points of nonlinear mappings, i.e. points at which the standard invertibility of the linearization breaks down. The reason for this is that by the very definition of Fredholm, the dimension at which the linearization breaks down must be finite dimensional. Thus we make the following

Definition   A C^1 Fredholm operator A acting between two spaces X and Y is <u>regular</u> at a point x if $A'(x)$ is a surjective linear mapping. If s is not regular the point x is called a <u>singular</u> point. The images of a regular and singular point are called <u>regular</u> and <u>singular values</u>, respectively.

Here are some examples of nonlinear Fredholm operators of index 0 acting between Banach spaces that we will meet in this text.

i) The periodic Riccati operator acting between the T-periodic function of $W_{1,2}(0, T) \rightarrow L_2(0, T)$.

ii) The periodic operator A defined by the formula $A(x) = x' + P_N(x)$ with $P_N(x)$ any real polynomial of degree N in x, acting between the spaces of (i).

Appendix 2   Bifurcation from Equilibria for Certain Infinite-Dimensional Dynamical Systems

In this appendix we consider the bifurcation of T-periodic solutions of the parabolic boundary-value problem

$$(1) \qquad \begin{cases} \dfrac{\partial u}{\partial t} + Lu + f(x, u) = g(x, t) \ \ \text{in} \ \ \Omega \end{cases}$$

$$u|_{\partial \Omega} = 0$$

obtained for a given T-periodic smooth forcing term $g(x, t)$ in (1). Here $x \in \Omega$ an arbitrary bounded domain in $\mathfrak{R}^N$ with boundary $\partial \Omega$. Moreover, L is a uniformly elliptic formally self-adjoint second order differential operator defined on Ω. The smooth function $f(u, x) \in \mathfrak{R} \times \Omega$ is specialized for the present to be convex with f_u bounded and to have the asymptotic properties

$$\lambda_2 > \lim_{t \to \infty} \frac{f(t, x)}{t} > \lambda_1 \ ; \ \ \lambda_1 > \lim_{t \to -\infty} \frac{f(t, x)}{t} > 0$$

where λ_1 and λ_2 are the lowest two eigenvalues of L relative to Ω.

The bifurcation phenomena reported here differs substantially from the Hopf bifurcation generally discussed for the change from equilibria to T-periodic solutions. It represents an extension to time-dependent from steady problems of the bifurcation mathematical structures recently discussed by me relative to infinite-dimensional Whitney fold and cusp singularities.

§1. The Static Case

Consider the special case, in which the forcing function g in (1) does not depend on t. Then the periodic solutions of (1) also do not depend on t, so (1) reduces to the nonlinear elliptic Dirichlet problem

(2) $Lu + f(x, u) = g(x)$ in Ω

$$u|_{\partial\Omega} = 0$$

This problem has been studied extensively, (see the above Section 1.4).

In terms of functional analysis, it suffices to consider weak solutions of (2) (for $g \in L_2(\Omega)$) in the Sobolev space $H = \overset{\circ}{W}_{1,2}(\Omega)$. Then the bifurcation points of (2) coincide with the singular points of the operator $A : Lu + f(x, u) : H \to H$ defined via duality, see above.

This operator A turns out to be a nonlinear Fredholm operator of index zero whose singular points have the following two properties:

(a) The singular points of A, S(A) form a connected infinite-dimensional manifold M. In fact M is a hypersurface in H of codimension one.

(b)   All the singular points of A are infinite-dimensional folds.

Thus one is lead to inquire: What happens to these results for the time-dependent case (1)?

§2. The Time-Dependent Periodic Case

We now consider the full equation (1) with the forcing term $g = g(x, t)$ depending explicitly on t in a T-periodic manner. The operator

$$(3) \qquad Bu = \frac{\partial u}{\partial t} + Lu = f(x, u) \qquad u|_{\partial\Omega} = 0$$

comprising the left hand side of (2) can be regarded as a smooth mapping between real Hilbert spaces X and Y of T-periodic function in t with

$$X = W_{1,2}[(0, T), H]$$

and

$$Y = L_2[(0, T), H]$$

with the usual norms.

In these terms we can prove the following extensions of our discussion of §1.

Theorem 1   The operator B regarded as a C^1 mapping between the real Hilbert spaces X and Y is a nonlinear Fredholm operator of index zero.

Theorem 2   Regular points of the mapping A are regular points of B. Moreover singular points of A are singular points of B and, in fact, for any $u \in H$

$$(4) \qquad \text{dimker } A'(u) = \text{dimker } B'(u)$$

Theorem 3   Any $u \in H \cap S(A)$ is an (infinite-dimensional) Whitney fold for B, provided B is regarded as a mapping between X and Y.

Theorem 4   For $g(x, t)$ smooth, restricted to a small neighborhood of a singular value of A in Y, equation (1) has exactly 2, 1 or 0 smooth, real T-periodic solutions in an appropriate neighborhood of the associated singular point in X.

In other words, bifurcation from equilibrium for T-periodic solutions of (1) occurs precisely at singular points u   at   A. Moreover the fold-type bifurcations occurring for the static problem go over to the same fold bifurcations for the periodic problem.

§3.   Sketch of the Proofs

On the Proof of Theorem 1   The operator Bu acting between X and Y can be written as the sum of the invertible linear operator $B_0 = u_t + Lu$ plus a compact operator, as follows from the Sobolev-Kondrachov results. Here

$$(5) \qquad\qquad B_0^{-1} w = e^{-tL}(I - e^{-TL})^{-1} Kw(T) + Kw(t)$$

$$\text{where } Kw(t) = \int_0^t e^{(s-t)L} w(s) ds$$

On the Proof of Theorem 2   Suppose (based on Theorem 1) that for $v(x, t) \in \text{Ker } B'(u)$

(6)
$$v(x, t) = \sum_{i=1}^{v} C_k(t)v_k(x) + w(x, t)$$

where $w \perp v_k$ for all t, k and $v_1, v_2 \ldots v_v$ form an orthonormal basis for Ker $A'(u)$. We prove C_k is independent of t and $w(x, t) \equiv 0$. The result establishes equation (4), and so the full result.

On the Proof of Theorem 3   This follows from our characterization of an infinite-dimensional Whitney fold between X and Y described above. This requires, for example, dimker $A'(u) = 1$ as established above for any singular point u of A. So that via (3) dimker $B'(u) = 1$. The second requirement for a fold follows from the properties of f described at the beginning of this note, especially the convexity of f in u.

On the Proof of Theorem 4   A consequence of parabolic regularity theory for this problem and the canonical normal form for an infinite-dimensional fold singularity .

§4. Extensions

The analysis given here can be directly extended to more equations and systems than (1). For example L can be of order 2m and $f(u)$ need not be bounded. The associated bifurcation or convex analysis must be considerably extended to cover Whitney cusps as well as more general singularities.

Appendix 3 - Elementary Facts About the Linear Dirichlet Problem

Throughout this section G is a bounded domain in $\mathfrak{R}^n$ with boundary ∂G and $x = (x_1, x_2, \ldots, x_n)$ denotes a general point of G. We consider the Laplace operator $\Delta = \Sigma_{i=1}^{n} \dfrac{\partial^2}{\partial x_i^2}$ defined over G.

Moreover throughout this appendix and in fact, through the entire book we use the elementary facts about Hilbert space found in the Problem Collection at the end of the book.

Problem 1 - The Linear Dirichlet Problem   Suppose that $g(x)$ is a function defined on G. Find a real-valued function $u(x)$ defined on $\bar{G}$ satisfying

$$\Delta u = g(x) \quad \text{on } G$$

$$u = 0 \quad \text{on } \partial G$$

Problem 2 - The Linear Eigenvalue Problem   Find all real numbers λ and real-valued functions $u(x)$ $(\neq 0)$ such that

$$\Delta u + \lambda u = 0 \quad \text{on } G$$

$$u = 0 \quad \text{on } \partial G$$

The Laplace operator Δ and the Dirichlet problem are the classic examples of elliptic operators and elliptic boundary value problems, respectively.

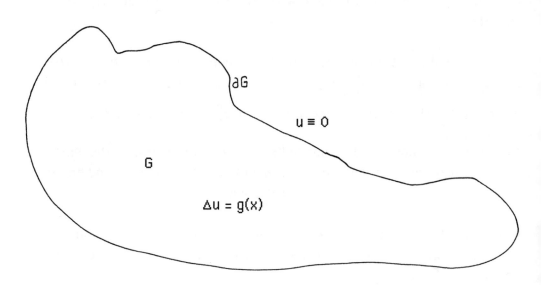

Illustrating the Linear Dirichlet Problem

A well-known reformulation of Problems 1 and 2 is to assign to each problem a so-called Green's function $K(x, y)$: where $K(x, y)$ satisfies

$$\Delta K(x, y) = \delta(x-y) \quad \text{for } x \in G$$

$$K(x, y) = 0 \quad \text{for } x \in \partial G$$

and for all $y \in G$. Here $\delta(x-y)$ denotes the Dirac delta function. Then Problems 1 and 2 can be written in integral equations as follows:

For <u>Problem</u> 1

$$u(x) = \int_G K(x, y)g(y)dy$$

For <u>Problem</u> 2

$$u(x) = \lambda \int_G K(x, y)u(y)dy$$

If we denote by $C_0(\overline{G})$ the Banach space of continuous functions $u(x)$ defined on $\overline{G}$, vanishing on ∂G, with norm $\|u(x)\| = \sup_G |u(x)|$, then it can be demonstrated that the linear mapping $Lf(y) = \int_G K(x, y)f(y)dy$ is a compact mapping of $C_0(\overline{G})$ (the set of continuous functions on $\overline{G}$, vanishing on ∂G) into itself.

Should the function $g(x)$ of Problem 1 be continuous but not necessarily Lipschitz continuous, then the associated Dirichlet Problem might not be solvable in the usual sense. Hence some other viewpoint is necessary.

Another approach to both Problems 1 and 2 is to divide each problem into two parts: an existence part, in which one proves that the problem has a solution in some "averaged" sense; and a regularity part, in which the above solution is proved to be sufficiently smooth, that is, to satisfy the equation at each point of G and the boundary condition. In order to settle the existence part of Problems 1 and 2 we define so-called weak solutions of both problems. These solutions are defined by means of the Dirichlet form

$$D[u, v] = \sum_{i=1}^{n} \int_G u_{x_i} v_{x_i}$$

associated with the Laplace operator Δu.

These averaged or "generalized" solutions are not necessarily differentiable in the classical sense. However, they can be regarded as limits of differentiable functions in the following sense:

Definition A function $u(x) \in L_2(G)$ possesses a square integrable generalized partial derivative u_{x_i} if there is a sequence of continuously differentiable functions $u_n(x)$ such that $u_n(x) \to u(x)$ and $(u_n(x))_{x_i} \to u_{x_i}$ in $L_2(G)$. (Note that the generalized derivative is unique (apart from a set of measure zero) if it exists).

The totality of functions $u(x) \in L_2(G)$ all of whose first order generalized partial derivatives are square integrable forms a linear space which we call $W_{1,2}(G)$. We make $W_{1,2}(G)$ into a Hilbert space by defining the inner product

$$(u, v)_{1,2} = \int_G uv + D[u, v]$$

This Hilbert space is an example of a Sobolev space.

The set of infinitely differentiable functions vanishing outside a compact subset of G is denoted $C_0^\infty(G)$; clearly $C_0^\infty(G) \subset W_{1,2}(G)$.

The closure of these functions in $W_{1,2}(G)$ forms a closed linear subspace which we denote by $\dot{W}_{1,2}(G)$.

Definition   A generalized solution of Problem 1 is a function $u(x) \in \dot{W}_{1,2}(G)$ such that

$$D[u, \phi] = \lambda \int_G u(x)\phi$$

for all functions $\phi \in \dot{W}_{1,2}(G)$.

Definition   A generalized solution of Problem 2 is a function $u(x) \in \dot{W}_{1,2}(G)$ satisfying

$$D[u, \phi] = \lambda \int_G u(x)\phi$$

for some fixed λ and all $\phi \in \dot{W}_{1,2}(G)$.

The term "generalized" solution is justified by the following result.

Lemma   Smooth solutions of Problems 1 and 2 are generalized solutions.

Proof   Suppose $u(x)$ is a twice continuously differentiable solution of Problem 1. Let $\phi \in \dot{W}_{1,2}(G)$. Then multiplying

$$\phi \Delta u = g(x)\phi$$

and integrating over G

$$\int_G \phi \Delta u = \int_G g(x)\phi$$

Integrating by parts and using the fact that u vanishes on ∂G we obtain

$$D[u, \phi] = -\int_G g(x)\phi$$

Thus u is a generalized solution of Problem 1 as u is automatically an element of $\overset{\circ}{W}_{1,2}(G)$.

The proof for Problem 2 is analogous.

The converse of this lemma is the regularity part of the Dirichlet problem mentioned above. The interested reader can find a complete discussion of regularity in the monograph of Gilbarg and Trudinger, Elliptic Partial Differential Equations of the Second Order, Springer-Verlag.

For the remainder of this section the following notation and basic results will be denoted $\|u\|_{1,2}$. If $v \in L_2(G)$ is norm is denoted $\|v\|_{0,2}$. Thus $\|u\|_{1,2}^2 = D[u] + \|u\|_{0,2}^2$ where $D[u] = D[u, u]$. Henceforth all inner products refer to $\overset{\circ}{W}_{1,2}(G)$, and we shall denote them simply by $(\, , \,)$. Furthermore, as G is a bounded domain in $\mathfrak{R}^n$, the following facts will be used:

a) (Poincaré's Inequality). For any $u \in \overset{\circ}{W}_{1,2}(G)$, $\|u\|_{0,2}^2 \leq KD[u]$, where K is a positive constant independent of u.

b) For any $u \in \overset{\circ}{W}_{1,2}(G)$, $D[u] = 0$ if and only if $u \equiv 0$.

c)   (Rellich-Kondrachev lemma).   If $u_n \to u$ weakly in $\dot{W}_{1,2}(G)$, then $u_n \to u$ strongly in $L_2(G)$ and more generally $u_n \to u$ strongly in $L_p(G)$ for $p < 2N/(N-2)$ [in case $N = 2$, the result holds for any $p < \infty$].

d)   (Sobolev's inequality).   For any $u \in \dot{W}_{1,2}(G)$,   and $p \leq 2N/(N-2)$ [in case $N = 2$, the result holds for any $p < \infty$].

$$\|u\|_{0,p}^2 \leq K_p D[u]$$

where K_p is a positive constant independent of u.   [in case $N = 2$, the result holds for any $p < \infty$].

e) Sobolev's Inequality and Kondrachev's Compactness Theorem for the Sobolev space $W_{1,2}(G)$. For a broad class of domains G, both results hold for the Sobolev space $W_{1,2}(G)$ with the Dirichlet integral replaced by $\|u\|_{1,2} = (\|u\|_{0,2}^2 + D(u))^{1/2}$. Thus in one dimension, the results hold for any finite interval, and in differential geometry the results hold for smooth compact Riemannian manifolds.

Theorem   If $g(x) \in L_2(G)$, the generalized solution of Problem 1 exists and is unique.

Proof   First we show that the generalized solutions of Problem 1 can be identified with the solutions of an operator equation $Lu = f$ in $\dot{W}_{1,2}(G)$. This equation is obtained as follows: $D[u, \phi]$ is a continuous bilinear functional in ϕ and u on the Hilbert space $\dot{W}_{1,2}(G)$. Thus by a standard result in Hilbert space theory, $D[u, \phi] = (Lu, \phi)_{1,2}$ where L is a continuous linear mapping of $\dot{W}_{1,2}(G)$ into itself. Also, $\int_G g(x)\phi$ is a linear functional in ϕ defined on $\dot{W}_{1,2}(G)$.

As $g(x) \in L_2(G)$, this functional is continuous. Indeed, let $\phi_n \to \phi$ strongly in $\dot{W}_{1,2}(G)$, then

$$\left| \int_G g(x)(\phi - \phi_n) \right| \leq \|g\|_{0,2} \|\phi - \phi_n\|_{0,2} \leq \|g\|_{0,2} \|\phi - \phi_n\|_{1,2}$$

Thus $\int_G g\phi_n \to \int_G g\phi$. Consequently, by the Riesz Representation Theorem, $\int_G g\phi = -(f, \phi)_{1,2}$ where $f \in \dot{W}_{1,2}(G)$. Thus if u is a generalized solution of Problem 1,

$$D[u, \phi] + \int_G g\phi = (Lu - f, \phi) = 0$$

for all $\phi \in \dot{W}_{1,2}(G)$. Hence $Lu = f$. Conversely, if $u \in \dot{W}_{1,2}(G)$ and satisfies $Lu = f$, then u is a generalized solution of Problem 1.

The second step then proceeds by noting that $Lu = \text{grad} \left(\frac{1}{2} D[u] \right)$ and $D[u]$ is weakly lower semi-continuous. This guarantees then that $Lu = f$ has a solution in $\dot{W}_{1,2}(G)$ provided $D[u]/\|u\| \to \infty$ as $\|u\| \to \infty$. That this last proviso holds can be seen as follows. By Poincaré's inequality [as in (a) above]

$$\|u\|_{0,2}^2 \leq KD[u]$$

Thus

$$\|u\|_{1,2}^2 = D[u] + \|u\|_{0,2}^2 \geq (1+K)D[u]$$

and

$$\frac{D[u]}{\|u\|_{1,2}} \geq \frac{1}{1+K}\|u\|_{1,2} \to \infty \quad \text{as} \quad \|u\|_{1,2} \to \infty$$

Finally, to prove uniqueness, we note that if u and v are generalized solutions of Problem 1, the difference $w = u - v$ satisfies $D[w, \phi] = 0$ for all $\phi \in \dot{W}_{1,2}(G)$. In particular, setting $\phi = w$, we have $D[w, w] = D[w] = 0$. Thus $w = 0$, that is $u = v$.

Theorem Problem 2 has a countably infinite number of distinct generalized solutions u_n with associated eigenvalues $\lambda_n \to \infty$.

Proof The existence of one solution u_1 with $\int_G u_1^2 = 1$ is a consequence of Theorem 3-15 of Section 3-3. The proof is carried out exactly as in Theorem 4-16 by translating the problem into the question of solvability of an operator equation in the Hilbert space $\dot{W}_{1,2}(G)$. In this case, we show the complete continuity of the operator $\mathfrak{I}_n \to \mathfrak{I}u$ and $\mathfrak{I}u$ is a completely continuous mapping. Furthermore, the reader will verify that $\mathfrak{I}u$ is the gradient of the functional $\frac{1}{2}\int_G u^2$ in $\dot{W}_{1,2}(G)$.

The linearity of Problem 2 can then be used to define a sequence of solutions $u_2, u_3, \ldots$ by applying the above mentioned Theorem 3-15 successively to the Hilbert spaces obtained by taking the orthogonal complement of u_1 in $\dot{W}_{1,2}(G)$, the orthogonal complement of the subspace formed by the vectors u_1, u_2, in $\mathring{W}_{1,2}(G)$ respectively. Then the countably infinite number of distinct solutions so constructed are associated with eigenvalues $\lambda_n \to \infty$.

We summarize the material in this appendix by the following diagram, which we call the "Solution Diamond." It shows, for the nonlinear elliptic problems discussed in this appendix, 4 equivalent formulations of the same problem. Thus, beginning at any side of the diamond, we can get to any other vertex and its reformulation in steps that are equivalent. The diagram can be written as follows:

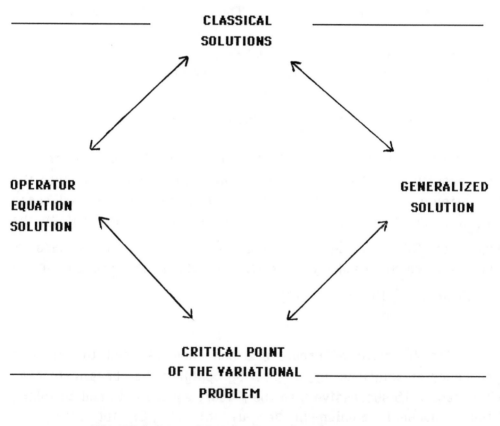

CLASSICAL SOLUTIONS

OPERATOR EQUATION SOLUTION

GENERALIZED SOLUTION

CRITICAL POINT OF THE VARIATIONAL PROBLEM

The Solution Diamond Diagram

Appendix 4 — On Besicovitch Almost Periodic Functions

Some results on the spaces of almost periodic functions in the sense of Besicovitch.

I. Definitions

(1) {u.a.p} = the set of all uniformly almost periodic functions, i.e. the closure of the trigonometric polynomials, in the supremum norm in the space continuous functions on the real line. These functions were studied carefully first by Harald Bohr.

(2) $M\{f(x)\} = \lim\limits_{T \to \infty} \dfrac{1}{2T} \int\limits_{-T}^{T} f(x)dx$ = the mean value of u.a.p. function f(x).

(3) $\overline{M}\{f(x)\} = \overline{\lim\limits_{T \to \infty}} \dfrac{1}{2T} \int\limits_{-T}^{T} f(x)dx$

(4) $C_B P\{u.a.p.\}$ = the closure of {u.a.p.} under the norm

$$\|f\|_{B^P}^{P} = \overline{M}\{|f(x)|^P\} = \overline{\lim\limits_{T \to \infty}} \dfrac{1}{2T} \int\limits_{-T}^{T} |f(x)|^P dx \qquad (*)$$

(5) A = the set of all trigonometric polynomials of form

$$S(x) = \Sigma a_j e^{i\lambda_j x}$$

where the coefficients a_j, and the exponents λ_j are real numbers and the sum contains only a finite number of terms.

(6) $C_{BP}(A)$ = the closure of A under the norm (*)
      $C_B^2(A)$ = { Besicovitch almost periodic function }
      i.e.

      If f(x) is Besicovitch almost periodic, then
      (1) f(x) is measurable
      (2) for any $\varepsilon > 0$. there is a trigonometric polynomial $P_\varepsilon(x)$,
      such that

$$\overline{M}\left\{\left|f(x)-P_\varepsilon(x)\right|^2\right\} = \varlimsup_{T\to\infty} \frac{1}{2T}\int_{-T}^{T}\left|f(x)-P_\varepsilon(x)\right|^2 dx < \varepsilon$$

(7)   f(x) is B^p almost periodic function if
      1) f(x) is $L_{loc}^p(\mathbb{R})$;
      2) to any $\varepsilon > 0$ corresponds a set of numbers

$$\dots\tau_{-2} < \tau_{-1} < \tau_0 < \tau_1 < \tau_2 < \dots$$

      such that
      i)   there is $l > 0$, such that

$$2 > \frac{\max\limits_{x\in(-\infty,\infty)} \text{ the number of terms in } \left[\{\tau_0,\tau_{\pm1},\tau_{\pm2}\dots\}\cap[x,x+1]\right]}{\min\limits_{x\in(-\infty,\infty)} \text{ the number of terms in } \left[\{\tau_0,\tau_{\pm1},\tau_{\pm2},\dots\}\cap[x,x+1]\right]}$$

      ii)   for each i,

$$\overline{M}_x\left\{\left|f(x+\tau_i) - f(x)\right|^p\right\} = \varlimsup_{T\to\infty}\frac{1}{2T}\int_{-T}^{T}\left|f(x+\tau_i) - f(x)\right|^p dx < \varepsilon_3^p$$

(iii) for every $C > 0$,

$$\overline{M}_x \overline{M}_i \frac{1}{C} \int_x^{x+C} |f(t+\tau_i) - f(t)|^P dt$$

$$= \lim_{T \to \infty} \frac{1}{2T} \int_{-T}^{T} \{ \lim_{n \to \infty} \frac{1}{2n+1} \Sigma_{i=-n}^n \frac{1}{C} \int_x^{x+C} |f(t+\tau_i) - f(t)|^P dt \} dx < \varepsilon^P.$$

II. Besicovitch's Results

1) C_B^p {u.a.p.} is a linear space.
2) If $f(x)$ is an element of $C_B(A)$, then $M\{f(x)\}$ exists.
3) $C_B^p (A) = C_B^p$ {u.a.p.} = {B^p almost periodic},

   In particular,
   $C_B^2(A) = C_B^2$ {u.a.p.} = { B^2 almost periodic} = {Besicovitch a.p.}

III. We will **prove** that

I) If $f(x), g(x) \in C_B^2(A)$, then $f(x)g(x) \in C_B(A)$. Thus $M\{f(x)g(x)\}$
   exists.

II) $C_B^2(A)$ = the closure of A under the norm

$$\| f \|^2_{B^2} = M\{|f(x)|^2\} = \lim_{T \to \infty} \frac{1}{2T} \int_{-T}^{T} |f(x)|^2 dx$$

which is a Hilbert space with the inner product

$$(f,g)_{B^2} = \lim_{T \to \infty} \frac{1}{2T} \int_{-T}^{T} f(x)g(x)dx$$

III. If $f(x) \in C_0^\infty(\mathfrak{R})$, then $f(x) \in \{B^2$ almost periodic$\}$ and $\|f\|_{B^2} = 0$. Therefore, in the Hilbert space $C_{B^2}(A)$, every element is an equivalence class.

Proofs. (due to my former student Dr. Yi Ying Chen)
* Proof of (I): if $f(x), g(x) \in C_{B^2}(A)$, then $f(x)g(x) \in C_B(A)$.

Since $f(x), g(x) \in C_{B^2}(A)$, there are sequences $\{P_n(t)\}$ and $\{Q_n(t)\}$, such that

$$\lim_{n \to \infty} \overline{M}\left\{\left|f(x) - P_n(x)\right|^2\right\} = 0,$$

$$\lim_{n \to \infty} \overline{M}\left\{\left|g(x) - Q_n(x)\right|^2\right\} = 0.$$

Consequently, there is $C > 0$ such that

$$\overline{M}\left\{\left|g(x)\right|^2\right\} \le \overline{M}\left\{\left|g(x) - Q_1(x)\right|^2\right\} + \overline{M}\left\{\left|Q_1(x)\right|^2\right\} \le C,$$

$$\overline{M}\left\{\left|P_n(x)\right|^2\right\} \le \overline{M}\left\{\left|P_n(x) - f(x)\right|^2\right\} + \overline{M}\left\{\left|f(x) - P_1(x)\right|^2\right\} + \overline{M}\left\{\left|P_1(x)\right|^2\right\} \le C.$$

Therefore,

$$\overline{M}\left\{\left|f(x)g(x) - P_n(x)Q_n(x)\right|\right\} \le \overline{M}\left\{\left|f(x) - P_n(x)\right||g(x)| + |g(x) - Q_n(x)||P_n(x)|\right\}$$

$$\le \overline{\lim_{T \to \infty}} \left(\frac{1}{2T}\int_{-T}^{T}\left|f(x) - P_n(x)\right|^2 dx\right)^{\frac{1}{2}} \overline{\lim_{T \to \infty}} \left(\frac{1}{2T}\int_{-T}^{T}\left|g(x)\right|^2 dx\right)^{\frac{1}{2}}$$

$$+ \varlimsup_{T\to\infty}\left(\frac{1}{2T}\int_{-T}^{T}|g(x)-Q_n(x)|^2 dx\right)^{\frac{1}{2}} \varlimsup_{T\to\infty}\left(\frac{1}{2T}\int_{-T}^{T}|P_n(x)|^2 dx\right)^{\frac{1}{2}}$$

$$\to 0 \text{ as } n\to\infty,$$

which shows $f(x), g(x) \in C_B(A)$.

* Proof of (II): $C_{B^2}(A)$ = the closure of A under the norm $\|f\|_{B^2}^2 = M\{|f(x)|^2\}$.

By definition, $C_{B^2}(A)$ = the closure of A under the norm $\|f\|_{B^2}^2 = M\{|f(x)|^2\}$. That is, for any $f(x) \in C_{B^2}(A)$, there is a sequence of trigonometric polynomials $P_n(x)$ such that $\lim_{n\to\infty} \overline{M}\{|f(x)-P_n(x)|^2\} = 0$. According to (I), $|f(x)-P_n(x)|^2 \in C_B(A)$. Thus $M\{|f(x)-P_n(x)^2|\}$. Therefore,

$$\lim_{n\to\infty} M\{|f(x)-P_n(x)|^2\} = \lim_{n\to\infty} \overline{M}\{|f(x)-P_n(x)|^2\} = 0.$$

* Proof of (III): if $f \in C_0^\infty(\mathcal{R})$, then $f(x)$ is B^2 almost periodic with zero norm.

It is obvious that if $f \in C_0^\infty(\mathcal{R})$, then $\|f\|_{B^2}^2 = 0$.

We now show that if $f \in C_0^\infty(\mathcal{R})$, then the set of numbers $\{. . .,-1,0,1, . . .\}$ will satisfy the conditions (i) – (iii) in the definition (7).

First consider condition (i).

Choose $1 = 2$, then $\dfrac{\max}{\min} = \dfrac{3}{2} < 2$, (i) is satisfied.

Thus, (i) is satisfied. Now consider (ii).

Since $\overline{M}_x\left\{|f(x+i)-f(x)|^2\right\} \leq 2\left[\overline{M}_x\left\{|f(x+i)|^2\right\} + \overline{M}_x\left\{|f(x)|^2\right\}\right] = 0,$, (ii) has been established.

Finally, consider (iii).

Since $f \in C_0^\infty(\mathbb{R})$, for every $C > 0$, there is $N > 0$, such that $\displaystyle\int_x^{x+c} |f(t)|^2 dt = 0$ whenever $|x| \geq N$.

Thus,

$$0 \leq \frac{1}{2T}\int_{-T}^{T}\left\{\varliminf_{n\to\infty} \frac{1}{2n+1} \Sigma_{i=-n}^{n} \frac{1}{c}\int_x^{x+c} |f(t)|^2 dt\right\} dx$$

$$\leq \frac{1}{2T}\int_{-N}^{N}\left\{\varliminf_{n\to\infty} \frac{1}{2n+1} \Sigma_{i=-n}^{n} \frac{1}{c}\int_x^{x+c} |f(t)|^2 dt\right\} dx$$

$$\leq \frac{1}{2T} 2N \varliminf_{n\to\infty} \frac{2n+1}{2n+1} \frac{1}{c}\int_{-\infty}^{\infty} |f(x)|^2 dt \to 0 \text{ as } T\to\infty.$$

On the other hand, for fixed x, there exists $\tilde{N} > 0$, such that if $|i| \geq \tilde{N}$, then

$$f(t+i) \equiv 0 \quad \text{for} \quad t \in (x, x+c).$$

Thus,

$$\overline{\lim_{n \to \infty}} \frac{1}{2n+1} \Sigma_{i=-n}^{n} \frac{1}{c} \int_{x}^{x+c} |f(t+i)|^2 dt = \overline{\lim_{n \to \infty}} \frac{1}{2n+1} \Sigma_{i=-N}^{N} \frac{1}{c} \int_{x}^{x+c} |f(t+i)|^2 dt$$

$$\leq \overline{\lim_{n \to \infty}} \frac{1}{2n+1} (2\tilde{N}+1) \frac{1}{c} \int_{-\infty}^{\infty} |f(x)|^2 dx = 0.$$

Therefore,

$$0 \leq \overline{M}_x \overline{M}_i \frac{1}{c} \int_{x}^{x+c} |f(t+i)-f(t)|^2 dt \leq 2\overline{M}_x \overline{M}_i \left\{ \frac{1}{c} \int_{x}^{x+c} \left(|f(t+i)|^2 + |f(t)|^2 \right) dt \right\} = 0,$$

(iii) has been obtained.

Chapter 2 – General Principles for Nonlinear Systems

Although nonlinear systems vary a great deal throughout science, their study is unified by fundamental mathematical principles. In this section we outline the simplest of these principles and illustrate just how they arise in practice. Of course, we saw in Chapter 1 how certain abstract principles were used in studying the simplest nonlinear systems, namely, integrable systems and their perturbations. These principles have the virtue of great "robustness" in the sense that they also have great value in studying more complicated nonlinear systems. In fact, one of the goals of this chapter is to illustrate just how the principles discussed briefly in Chapter 1 can be extended to study more general nonlinear systems such as occur throughout science and technology. Such systems are generally nonintegrable, even with our extended definition.

The principles to be included in our discussion are:
- linearization,
- the calculus of nonlinear operators,
- nonlinear eigenvalue problems and parameter dependence for nonlinear problems;
- iteration, and how these processes leads to both bifurcation and chaotic dynamics,
- variational principles for the characterization of the solutions to nonlinear problems,
- bifurcation and its relationship to the idea of singularities of nonlinear mappings,
- the idea of steepest descent to locate the absolute minimum of a functional,
- the function spaces of Sobolev to adapt generalized derivatives into a Hilbert space context,

– nonlinear Fredholm operators to describe general operators in science that have a coherent mathematical structure relative to the abstract constructions in this book.

Section 2.1 – Differentiable Nonlinear Operators

In the sequel we shall often discuss the properties of a nonlinear operator A acting between infinite dimensional spaces X and Y. We write

$$A : X \to Y$$

Generally we shall suppose that the spaces X and Y are linear spaces and (in addition) complete metric spaces (X,d) (Y,d_1) whose metrics are given by norms, i.e.

$$d(x, \check{x}) = \|x - \check{x}\|_X$$

and

$$d_1(y, \bar{y}) = \|y - \bar{y}\|_Y$$

Such spaces X and Y are called Banach spaces commemorating the contributions to this subject of the great Polish mathematician Stefan Banach.

Given such an operator A acting between Banach spaces X and Y, it is important to define the linearization of A at a given point $x \in X$. To this end the concept of derivative of A at x is important. We shall see A is differentiable at x with derivative L = A'(x) if there is a bounded linear operator L

$$\| A(x+h) - A(x) - Lh \|_y = o(\| h \|)$$

for all sufficiently small $h \in X$ as $\| h \| \rightarrow 0$.

Simple Cases

(i) Suppose $A : \Re^n \rightarrow \Re^n$ is a C' differentiable vector - valued function defined on $\Re$. Thus setting $X = Y = \Re^n$ we find A'(x) = the Jacobian matrix of A.

(ii) Suppose $L : \Re^N \rightarrow \Re^n$ is a n x n matrix of real constants. Proceeding as above we find L'(x) = L for all x .

Both of the above cases were finite dimensional, i.e. the spaces X and Y were $\Re^n$. Now we list some examples where X and Y are infinite dimensional.

(iii) Suppose $A : (x_1, x_2 , x_3 ... x_n , ...) = (x_1^2 , x_2 , x_3 ...)$ is a mapping of $l_2 \rightarrow l_2$, the Banach space of real sequences with norm square equal to the square of the sum of the components of each sequence. Then the mapping A is differentiable with

$$A'(x)h = (2x_1 h_1, h_2 , h_3 , ... h_n ...)$$

In fact, the mapping $A'(x)$ is a linear Fredholm operator of index zero mapping of $l_2 \to l_2$ for each x, and thus is referred to as a nonlinear Fredholm operator of index zero (cf. Appendix 1 above). This map is called the infinite dimensional Whitney fold.

Clearly dim Ker $A'(0) = 1$, so that $x = 0$ is a singular point for the mapping. This mapping is the global normal form for an infinite dimensional fold and was used extensively in Chapter 1.

Nonlinear Eigenvalue Problems

In the sequel we shall often consider simple operator equations depending on a parameter of the form

(1) $u = \lambda N u$ $\| u \| = R$

defined on a real Hilbert space H, as R varies over the positive real numbers. Here

(i) λ is a real parameter the "nonlinear eigenvalue."
(ii) $u \in H$ is the "nonlinear eigenvector" and is required to be nonzero.
(iii) N is a general nonlinear operator mapping H into itself.

Let us begin with the simplest linear case and suppose N is a positive definite nonsingular self-adjoint $n \times n$ matrix A of real constants. Then the fundamental problem "<u>to find all the eigenvalues and eigenfunctions of (1)</u>" is solved by simple linear algebra. For the matrix A, there are n real eigenvalues written

$$\lambda_1 \geq \lambda_2 \geq \lambda_3 \ldots \geq \lambda_n$$

with associated orthonormal eigenvectors $u_1, u_2, u_3 \ldots u_n$. These vectors form a complete set in $\mathfrak{R}^n$.

The results are illustrated on the picture directly on the following graph, referred to in the sequel as a bifurcation diagram,:

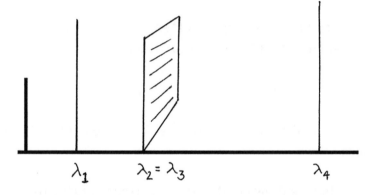

Figure 2.   Bifurcation Diagram ($\| u \|$, λ) plane for
              Lu = λu . (Here L has positive eigenvalues).

Figure 2.1 consists of n vertical lines commencing at $(0, \lambda_i)$ as i varies from 1 to n. When the eigenvalues are nondistinct (say $\lambda_2 = \lambda_3$), the two lines ($\| u \|$, λ_3) merge and form a two-dimensional plane.

The key abstract question we consider is what happens when the operator is no longer the linear operator L(u)   but some nonlinear operator N(u).   How does the bifurcation diagram 2.1 change?

In order to answer this question we shall reformulate (i) in several ways :

(i) **Geometric reformulation**   Notion of invariant directions. Consider the unit sphere $\Sigma = \{ \| u \| = 1 \}$ in H . The operator N takes an element u on Σ to an image element Nu . Then equation (1) is equivalent to the problem   of invariant directions for a nonlinear operator N, i.e. find "invariant directions for N on Σ." For more details, see problems below.

(ii) **Calculus of Variations reformulation** Suppose N is a gradient operator defined on a Hilbert space H,   i.e. Nu = grad $\mathcal{N}$(u) where $\mathcal{N}$ is a C' real-valued function on H , in symbols,

$$\lim_{\varepsilon \to 0} \frac{\mathcal{N}\,(u + \varepsilon v) - \mathcal{N}\,(u)}{\varepsilon} = (Nu, v)_H \quad \text{for all } v \text{ in } H$$

Then (1) is equivalent to the problem : find the critical points of the function $\mathcal{N}$(u) on the unit sphere $\Sigma = \{ u \mid \| u \|^2 = R , \text{ a real}$ number $\}$. Then the nonlinear eigenvalue $\lambda = \lambda(R)$ is an associated Lagrange multiplier and is not prescribed, but generally varies as R runs over the real numbers, except of course, when the operator N is linear and self-adjoint.   In the simplest case mentioned above when H = $\mathbb{R}^n$ and A is a self-adjoint positive definite matrix, the associated functional is simply the quadratic form $\frac{1}{2}$ (Ax, x).

Nonlinear eigenvalue problems have already been encountered in Chapter 1.   In Section 1.7, periodic solutions of second order conservative systems of the form

$$\ddot{\underline{x}} + \lambda^2 \nabla V(\underline{x}) = 0$$

were discussed. A parameter λ measuring the period of solutions was introduced. Indeed, this parameter had to be introduced explicitly because it was not mentioned in the actual problem of finding global families of periodic motions for the conservative systems involved. Moreover, the periodic solutions of this equation were characterized by extremal principles. It is this viewpoint that leads to the extensions we now discuss.

We now extend this point of view to a general context in an infinite dimensional Hilbert space H. To this end, we begin by first examining the weak topology of H and then extending the notion of a bounded self-adjoint linear operator L to a nonlinear context. The reason we study weak convergence is that it is an interesting substitute in an infinite dimensional context for the usual notion of convergence in finite dimensions, especially when notions of compactness arise. The usual notion of convergence in an infinite dimensional Hilbert space, namely, norm convergence (also called strong convergence) suffers from the defect that bounded sets may not have a convergent subsequence. Thus, even in the linear case a bounded self-adjoint linear operator L may have no eigenvalues. To overcome these difficulties, it is useful to study the following facts.

<u>Weak</u> <u>convergence</u> in a Hilbert space H: by $x_n \to y$ weakly we mean, in terms of bounded linear functionals defined on H, $(x_n, z) \to (y, z)$ for all $z \in H$. In contrast, convergence in norm (or strong convergence) of elements $x_n \to y$ in H means $\|x_n - y\| \to 0$. In this regard we note the following facts: (a) weak limits are unique; (b) if $x_n \to y$ weakly in H, then the set of H-norms of the

sequence $\{x_n\}$ is uniformly bounded; (c) any sequence in H whose norms are uniformly bounded has a weakly convergent subsequence; (d) if $\|x_n - y\| \to 0$ (that is, $x_n \to y$ strongly), then $x_n \to y$ weakly; but not conversely; (e) in a finite dimensional Hilbert space weak and strong convergence coincide; (f) if $x_n \to y$ weakly in H, $\|y\| \le \underline{\lim} \|x_n\|$; (g) if $x_n \to y$ weakly in H, and $\|x_n\| \to \|y\|$, then $x_n \to y$ strongly in H.

Next we extend the notion of a continuous self-adjoint linear operator to a nonlinear context. Let f be a continuous mapping of a Hilbert space H into itself. Then f is a gradient operator provided there is a C^1 real-valued continuous functional F(u) defined on H, that is

$$\lim_{\varepsilon \to 0} [F(u + \varepsilon v) - F(u)]/\varepsilon = (f(u), v)$$

for all $u, v \in H$; and we write grad F(u) = f(u). If f(u) = 0, the element u is called a critical point of the functional F(v). Just as in the finite dimensional case, local maxima and minima of F(v) are critical points.

An alternative definition is as follows.

Lemma   Let f be a continuous mapping of a Hilbert space H into itself. Then f is a gradient operator if and only if the following identity holds for all $u, v \in H$

$$(*) \qquad \int_0^1 [(u, f(su)) - (v, f(sv))]ds = \int_0^1 (u - v, f(su + (1 - s)v))ds$$

Moreover, $F(u)$ can be written as $\int_0^1 (u, f(su))ds$.

Proof   If $F(u) = \int_0^1 (u, f(su))ds$ and (*) holds, then

$$F(u + \varepsilon v) - F(u) = \varepsilon \int_0^1 (v, f(u + s\varepsilon v))ds.$$

Let $\varepsilon \to 0$, then

$$\lim_{\varepsilon \to 0} [F(u + \varepsilon v) - F(u)]/\varepsilon = (v, f(u))$$

Conversely, if f is a gradient operator, then from the line above

$$(d/ds)F(u + sv) = \lim_{\varepsilon \to 0}[F(u+(s+\varepsilon)v) - F(u + sv)]/\varepsilon = (v, f(u + sv)).$$

Replacing u by v and v by $u - v$ and integrating from 0 to 1 with respect to s, we obtain the equation

$$(**) \qquad F(u) - F(v) = \int_0^1 (u - v, f(v + s(u - v))ds$$

and setting $v = 0$ and $F(0) = 0$, $F(u) = \int_0^1 (u, f(su))ds$.

Remark   This result is quite significant from the viewpoint of calculus of variations. It relates to the so-called inverse problem of that subject, by which we mean, suppose one is given a system of equations $f(u) = 0$ in a finite Hilbert space H; then how does one go about constructing a real-valued functional $F(u)$ whose "Euler-Lagrange equations" are the original system $f(u) = 0$. The formula given immediately above this remark shows exactly how this can be done, provided the operator f(u) that maps the Hilbert space into itself is a gradient operator. To see how this can be checked, we mention merely that assuming the operator is itself a C^1 mapping, one needs merely check that the Frechet derivative of f'(u) at each point u, is self-adjoint. See Problems.

An operator f is <u>completely</u> <u>continuous</u> of f maps weakly convergent sequences into strongly convergent sequences. In a Hilbert space we note that complete continuity implies compactness and continuity.

A functional F(u) is <u>continuous</u> <u>with</u> <u>respect</u> <u>to</u> <u>weak</u> <u>convergence</u> if $u_n \to u$ weakly implies $F(u_n) \to F(u)$. This insures that on any bounded set, F is bounded.

Lemma   If the gradient of a functional F(u) is completely continuous, F(u) is continuous with respect to weak convergence.

Proof Let $u_n \to u$ weakly and denote grad F(u) by f(u), then by virtue of (**)

$$F(u_n) - F(u) = \int_0^1 (u_n - u, f(u))ds + \int_0^1 (u_n - u, f(u + s(u_n - u)) - f(u))ds.$$

The first term tends to 0 as n tends to ∞, by weak convergence. The second term also tends to 0. Indeed, by the complete continuity of f, $\|f(u + s(u_n - u)) - f(u)\| \to 0$; and $\|u_n - u\|$ is uniformly bounded.

We now consider a simple nonlinear eigenvalue problem, and prove

Theorem   Suppose that $f(u) = \operatorname{grad} F(u)$ is a completely continuous gradient operator mapping a Hilbert space H into itself with

  (i) $f(0) = 0$ and
  (ii) $(f(u), u)_H > 0$ for $u \neq 0$.

Then for every positive number c there is an element u_c on the sphere $\Sigma_c = \{u | \|u\|^2 = c\}$ satisfying the equation

(1) $u_c = \lambda f(u_c)$

where $\lambda = c(f(u_c), u_c)^{-1}$

Here u_c is characterized as the maximum of the functional $F(u)$ on the sphere Σ_c for every real number c.

Proof   The proof is divided in two parts. In Part 1 we find an extremal element u_c satisfying

$$F(u_c) = \max_{\Sigma_c} F(u).$$

In Part 2, we show u_c is a critical point of the functional $F(u)$ on the sphere Σ_c and so satisfies the equation (1).

Part 1   Clearly $\alpha = \sup_{\Sigma_c} F(u)$ is a finite number. Let the sequence $u_n \in \Sigma_c$ be chosen so that $F(u_n) \to \alpha$. Clearly the sequence $\{u_n\}$ is bounded in norm and consequently by the properties of weak convergence mentioned above $\{u_n\}$ has a weakly convergent subsequence also relabelled $\{u_n\}$ so that $u_n \to \bar{u}$ weakly in H. By the complete continuity of $f(u) = \operatorname{grad} F(u)$ and the resulting weak continuity of the functional $F(u)$, $F(\bar{u}) = \alpha$. However, there is the possibility that $\|\bar{u}\|^2 < c$ since by the properties of weak convergence mentioned earlier, at this stage one knows only that $\|\bar{u}\|^2 \leq c$. To ensure that $\|\bar{u}\|^2 = c$ and so to conclude $\bar{u} = u_c$ we note that if $\bar{u}$ has norm smaller than expected, the ray tu_c for $t > 0$ will intersect Σ_c for some $t_0 > 1$. Consequently, the short computation below

$$F(t_0\bar{u}) - F(\bar{u}) = (t_0 - 1) \int_0^1 (\bar{u}, \text{grad } F(k(s)u)ds$$

with $k = 1 + s(t_0 - 1) \geq 1$, shows we obtain a contradiction to the maximality of $F(\bar{u})$.

Part 2   Consider small arcs $u_c(t)$ on Σ_c centered at u_c of the form

$$u_c(t) = \left[\frac{u_c + ty}{\sqrt{c + t^2\|y\|^2}} \right]\sqrt{c}$$

for all directions y orthogonal to u_c in H. Clearly $\|u_c(t)\|^2 = c$, so that indeed, $u_c(t)$ is an arc on Σ_c. Moreover, a small computation using the result of equation (**) shows

$$F(u_c(t)) - F(u_c) = t(y, f(u_c)) + o(t)$$

Hence, by the extremal characterization of u_c, we find that for each element y satisfying the orthogonality relation

$$(y, u_c) = 0 \quad \text{it follows that} \quad (y, f(u_c)) = 0.$$

Indeed, otherwise, one would have a contradiction to the maximality of $F(u_c)$ on Σ_c.

Using the Hilbert space orthogonal decomposition property, the projection theorem implies that

$$f(u_c) = \lambda u_c + \bar{v} \qquad \text{where} \quad u_c \perp \bar{v}$$

Thus, taking the inner product with $\bar{v}$, we find from the last equation that

$$(f(u_c), \bar{v}) = \|\bar{v}\|^2$$

Now, by the choice of $\bar{v}$, we then find $\|\bar{v}\|^2 = 0$, which implies $\bar{v} = 0$. The formula for the eigenvalue λ is obtained merely by taking the inner product of the equation (1) with the extremal element u_c itself.

(iii) <u>An iteration scheme for</u> (1). We consider the following iteration scheme for the completely continuous operator N. As before, we assume $(Nu, u)_H > 0$ for $u \neq 0$. Start with a point u_0 on the unit sphere. Then $Nu_0 \neq 0$ by assumption, and one computes the first step in the iteration as follows:

$$u_1 = \lambda_1 N(u_0) \qquad\qquad \| u_0 \| = 1$$

One can always compute the number λ_1 by requiring that this positive number b so chosen that $\lambda_1 \|Nu_0\| = 1$. One then computes the second element in the iteration u_2 on the unit sphere and the positive number λ_2 as follows:

$$u_2 = \lambda_2 N(u_1) \qquad\qquad \| u_1 \| = 1$$

Continuing with this process, one constructs the sequence u_n by defining

$$u_{n+1} = \lambda_{n+1} N(u_n) \qquad \| u_n \| = 1$$

with the point u_n on the unit sphere where λ_{n+1} is determined from u_n by requiring that $\| u_{n+1} \| = 1$. Suppose that as $n \to \infty$, the sequence λ_n so constructed converges to $\lambda \neq 0$, and the sequence u_n converges weakly to v. Consequently, by the complete continuity of N, one finds that $N u_n$ converges strongly to Nv. Then the iteration scheme just described converges to a solution (v, λ) on the unit sphere for the eigenvalue problem $v = \lambda Nv$. It is interesting to find conditions on the operator N that guarantee the convergence of this iteration scheme.

One of the most important examples of a nonlinear eigenvalue problem occurs when one considers bifurcation processes from a linear eigenvalue problem. Thus for a completely continuous self-adjoint linear operator K in a Hilbert space, the results on the linear eigenvalue problem $\lambda u = Ku$ are well known, by the Riesz-Fredholm theory. However, if one adds Nu, a completely continuous C^1 higher order nonlinear operator, to K, one finds the new nonlinear eigenvalue problem

(2) $\lambda u = Ku + Nu$

For small $\| u \|$, it is natural to inquire just what effect the nonlinear term Nu has on the linear eigenvalue problem. To give easily visualized answers to this question, one constructs a "Bifurcation Diagram" already mentioned above. This diagram is a two-dimensional plot of solutions (u, λ) of (2) drawn on a graph with a λ axis versus a $\|u\|$ axis.

The general effect of the nonlinearity on the Bifurcation Diagram (in the simplest case when all the eigenvalues of the linear eigenvalue problem are simple) is that the straight lines $(\|u\|, \lambda_i)$ of the linear equation $\lambda u = Ku$ are deformed to the curves $(u(R), \lambda(R))$ as in the following diagram :

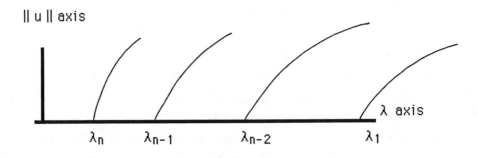

Figure 2. 2   **Bifurcation Diagram $(\|u\|, \lambda)$ plane for**
$Ku + Nu = \lambda u$. (Here we have assumed
Nu is a higher order completely continuous
gradient operator and the spectrum of K
consists of simple eigenvalues).

However, in realistic situations, many different cases can occur :

(i) the curves $C_i = (u(R), \lambda(R))$ may bend in a negative direction .

(ii) the curves C_i and C_j may cross for certain integers i and j .

(iii) (in case of multiple eigenvalues (i, i + 1) the curves C_i and C_{i+1} may interact and disappear. See examples.

Section 2.2   Iteration of Nonlinear Operators

An interesting nonlinear structure can be obtained by iteration of a given nonlinear operator A that maps a space X into itself. Simple examples for the space X would be the real axis $\mathcal{R}^1$, N dimensional Euclidean space $\mathcal{R}^N$, the complex numbers C^1, or in fact, any Banach space X. Arguing abstractly, one starts with a given point $x_0 \varepsilon X$, and then obtains successively the sequence

$$x_0, \ x_1 = Ax_0, \ \ x_2 = A^2 x_0, \ x_3 = A^3 x_0, \ldots$$

and, in the general case, after n steps

$$x_n = A^n x_0$$

One then inquires, What are the properties of the sequence $\{\, x_n \,\}$ so defined as $n \to \infty$? Does this sequence have any limit point $\overline{x}$? If so, what equation does the limit point satisfy? These questions are especially relevant for present day high speed computers are their associated computer graphics. Indeed, the field of chaotic dynamics can be thought of as a direct result of graphing iterations of nonlinear eigenvalue problems that depend crucially on the magnitude of relevant parameters.

In this connection the notion of fixed point is fundamental.

Definition   A fixed point of the mapping A in X is a point
   $x \in X$ that satisfies the equation $A(\overline{x}) = \overline{x}$.

To obtain such a fixed point for the mapping A in X assume for simplicity that the space X is a Banach space, and moreover, that

(i) $\lim\limits_{n\to\infty} x_n = A^n x_0$ exists

(ii) A is a continuous mapping of X into itself.

Then we claim

$$\bar{x} = \lim\limits_{n\to\infty} x_n = \lim\limits_{n\to\infty} A^n x_0$$

is a fixed point for the mapping A in X.

Proof $A(\bar{x}) = A(\lim\limits_{n\to\infty} x_n)$ by definition

$\qquad\qquad = \lim\limits_{n\to\infty} A(x_n)$ by continuity of A

$\qquad\qquad = \lim\limits_{n\to\infty} x_{n+1}$

$\qquad\qquad = \bar{x}$

Now we give a number of well-known examples in mathematical analysis where iteration processes yield interesting results.

Example A   First Order Linear Systems of Ordinary Differential Equations with Constant Coefficients

Consider attempting to solve by iteration the initial value problem

Mathematical Structures

(1)
$$\dot{x} = Ax \qquad x(0) = x_0, \qquad \dot{x} = \frac{dx}{dt}$$

where A is an $N \times N$ matrix of constant coefficients, and $x(t)$ is an N vector depending on t.

To solve this problem by iteration, one sets up a procedure in such a way that the solution of (1) can be obtained as a limit similar to the procedure described to obtain fixed points above.

Iteration Step

$$\text{Set} \quad x_0(t) = x_0$$

$$\dot{x}_1 = Ax_0 \qquad x_1(0) = x_0$$

$$\dot{x}_2 = Ax_1 \qquad x_2(0) = x_0$$

(2)
$$\dot{x}_{n+1} = Ax_n \qquad x_{n+1}(0) = x_0$$

One inquires: Does the iteration step lead to a solution of (1)? In fact it is known that the iteration step defined by (2) has the solution

(3)
$$x(t) = \exp(tA)x_0 \qquad \text{where } \exp A \text{ denotes the exponential of the matrix } A$$

Theorem   The initial value problem (1) has one and only one solution $\bar{x}(t) = \lim_{n \to \infty} x_n(t)$ given by (3) above.   In this case the space X is the space of continuous N-vector functions on the interval $(+ \infty, - \infty)$ with the sup norm.

Example B   Suppose A is a mapping of the complete metric space (X, d) into itself satisfying the inequality

$$d(Ax, Ay) \leq \propto d(x,y)$$

for all $x, y \in X$ with $\propto$ an absolute positive constant strictly less than unity. Then, as defined above, A is a continuous mapping of X into itself, $\lim_{n \to \infty} x_n = \overline{x}$ exists. Thus, by the above reasoning, we find

Theorem   (The Contraction Mapping Theorem)   Under the above hypotheses, A has a fixed point $\overline{x} \in X$. This fixed point is unique.

The proof of this result is quite standard by this time. Thus we refer the reader to the books of the Bibliography

Example C   The Initial Value Problem for First Order Systems of Nonlinear Differential Equations   Consider solving by iteration the initial value problem for the single nonlinear differential equation

$$\dot{x} = f(t, x) \qquad x(0) = x_0$$

where $f(t,x)$ is continuous in t and satisfies a Lipschitz condition in x.

Theorem   The above initial value problem has a unique continuous solution $\overline{x}(t)$ on a sufficiently small interval $(-a, a)$. The solution can be calculated by iteration via the initial value problem

$$\dot{x}_{n+1} = f(t, x_n) \qquad x_{n+1}(0) = x_0$$

This result clearly extends directly to systems of first order equations.

Example D   The Feigenbaum Map

Let $f(x) = kx(1-x)$ for $0 \le x \le 1$ and k a constant $1 \le k \le 4$. Then we consider the iteration

(4) $x_{n+1} = kx_n(1 - x_n)$ $x_0 \in (0, 1)$

Clearly the sequence $\{ x_n \}$ lies in the unit interval for all n.

The fixed point $\bar{x} = 1 - \frac{1}{k}$ is well defined for all $k \in (0,4]$. However $\bar{x}$ is stable (according to the linear stability theory) only if $k \in (1, 3)$. For $k > 3$, $\bar{x}$ is unstable, and the stability of the fixed point is lost to the fixed points of the second generation map $f \circ f(x)$ of $f(x)$. At $k = 3$, there is a bifurcation of fixed points for the Feigenbaum map as well as an exchange of stability. More explicitly, the fixed points of the second generation map become stable for $k > 3$ whereas the fixed points $\bar{x}$ become unstable exactly when $k > 3$. These fixed points $\bar{x}$ are however stable when $k \in (1, 3)$. This is the idea of exchange of stability, for if k increases beyond 3, the fixed points $\bar{x}$ loose stability to the fixed points of the second generation map. Here, stability means linearized stability at a fixed point, so for the fixed point $\bar{x}$, the criterion for stability is that for the map f, satisfying the iteration $x_{n+1} = f(x_n)$, $|f'(\bar{x})| < 1$, and instability means that $|f'(\bar{x})| > 1$. Notice here that $x = 0$ is a fixed point for all values of k, but is always excluded from the iteration procedures we discuss since it is unstable when $k > 1$. Indeed, $|f'(0)| = k$ for the Feigenbaum map. Also note that the Feigenbaum mapping $f(x) = \lambda x(1-x)$ maps the interval $[0, 1]$ into itself when λ varies between 1 and 4.

The process that triggers this exchange of stability for the Feigenbaum map is called <u>bifurcation</u>. This means that a single branch of fixed points existing for k less than a critical number splits into new families of fixed points when k exceeds this value. This process of splitting has been carefully studied mathematically and has great intrinsic interest.   The second generation map $f^{(2)}x$ obtained by iterating the Feigenbaum map twice, has, for each k , a new fixed point $x^{(2)}$ called a 2 cycle of numbers p, q   that differs from the fixed points of f itself.   This 2 cycle p, q   has the property that $p \neq q$, but $f(p) = q$ and $f(q) = p$. In all cases p,   q and the two fixed points of the Feigenbaum mapping f itself satisfy the same quartic equation. The roots of this equation all become real when $k \geq 3$ and the 2 cycle p, q gains the stability property when k is strictly greater than 3, but less than 3.4495. Moreover, as was mentioned above, $\bar{x}$, the fixed point of the mapping   f(x) = kx(1-x) becomes unstable when k > 3. As stated just above, this fixed point $\bar{x}$ loses its stability to the fixed points $x^{(2)}$ of the second generation map $f^{(2)}$.   Note that the $x^{(2)}$ are not fixed points of f(x) itself. For more information, see the Problems Section at the end of the book.

This process is called period-doubling bifurcation. Moreover, the 2 cycle p, q remains stable as k increases beyond 3 to 3.4495. Then the period 2 fixed points lose stability themselves at k = 3.4495 to a 4 cycle, i.e. period 4 fixed points of the Feigenbaum map that coexist with the fixed points of both this map and its second generation map.   Again, this process is known as period doubling bifurcation. Once more, at a number a_k in the interval (3, 3.5699) a 2^k cycle is born by a period doubling and bifurcation process k = 1, 2, 3, ... Feigenbaum noted that the numbers a_k tend to a limit a_∞ in a geometric progression with

$$a_k \text{ approximately } = a_\infty - cF^{-k}$$

Here, the following approximations hold:

$$a_\infty = 3.569946\ldots \qquad c = 2.6327\ldots \qquad F = 4.669\ldots$$

This last number is called the Feigenbaum constant and seems to occur in many different iteration problems. For this reason, it has a certain "universal" importance for nonlinear science. This process of period doubling and bifurcation is thus repeated to yield the so-called "Period Doubling Route to Chaos" for the Feigenbaum map. The same process is repeated for many other iterations. However, the analytical details of this process are still unclear at this writing.

It is interesting to note the case $k = 4$ for the Feigenbaum map (4). This case is special because it has an explicit solution even though the mapping itself is referred to as chaotic. This solution can be written

$$x_n = \sin^2(2^n \sin^{-1}\sqrt{x_0}) \qquad n = 1, 2, 3, \ldots$$

The reader is invited to verify this explicit formula, and to consider its meaning for numerical computation and graphics. One is led to ask, for example, when are iteration schemes, such as those defined by (4) above, integrable ?

A)   Steepest Descent Methods for Minimization of a Functional

It is possible to locate the absolute minimum of a real-valued function defined on all of Hilbert space H. Almost 150 years ago, A. Cauchy found a very flexible important process for this location. The process to be described below consists in solving the initial value problem for the system

(1) $\dfrac{dx}{dt} = -f(x)$ with grad $F = f$

$$x(0) = x_0$$

Here $F(x)$ is a C^2 real-valued function defined on a real Hilbert space H. The study of (1) and its limits as $t \to \infty$ is referred to as the method of steepest descent for locating the absolute minimum of F. Modifications of this method are used to study saddle points of the functional F, infinite-dimensional Morse theory, and critical point theory in general. These topics are left for future discussions.

One easily shows that (along a solution $x(t)$ of (1)), $F(x(t))$ decreases as $t \to \infty$. Provided the solution of (1) exists for all t, one attempts to show that $\lim\limits_{t \to \infty} x(t) = \bar{x}$ exists and is a solution of $f(x) = 0$. The convergence of the method is in question, and we now take up this problem.

If $F(x)$ possesses a strict relative minimum at some point x_∞, then the method of steepest descent is quite easily justified by the following

(2) **Theorem**   Suppose $F(x)$ is a C^2 real-valued functional defined on a sphere $S(x_0, r)$ of a Hilbert space H, and suppose that for some absolute constant $A > 0$

(3) $$(F''(x)y, y) \geq A \| y \|^2$$

for $x \in S(x_0, r)$ and $y \in H$.   Then   provided $\dfrac{\|F'(x_0)\|}{A} \leq r$, the initial value problem (1) has a unique solution defined for all t, $\lim\limits_{t \to \infty} x(t) = x_\infty$ exists and is the unique minimum of $F(x)$ in $S(x_0, r)$ as well as the unique solution of $f(x) = 0$ in $S(x_0, r)$. Furthermore, we have the following estimate for the rate of convergence of $x(t) \to x_\infty$,

(4) $$\| x(t) - x_\infty \| = O(e^{-At})$$

Proof   First we note that the initial value problem (1) has one and only one solution for small t by virtue of the local existence theory for ordinary differential equations. To ensure that $x(t)$ stays in $S(x_0, r)$ and that $\| \dfrac{dx}{dt} \| \to 0$ as $t \to \infty$ so that $\|f(x(t))\| \to 0$, we argue as follows. Along a solution $x(t)$ of (1),

(5) $$\frac{d}{dt} F(x(t)) = (f(x(t)), x'(t)) = - \| x'(t) \|^2$$

Hence $F(x(t))$ decreases as t increases. Also,

$$\frac{d^2}{dt^2} F(x(t)) = -2(x''(t), x'(t)) = 2(F''(x(t))x'(t), x'(t))$$

$$\geq 2A\|x'(t)\|^2 = -2A \frac{d}{dt} F(x(t))$$

Hence, as a function of t, $F(x(t)) = g(t)$ satisfies the second order differential inequality

$$g'' + 2Ag' \geq 0.$$

Consequently, by (5),

$$(\frac{d}{dt})F(x(t)) \geq -\|f(x_0)\|^2 e^{-2At}$$

and so

(6)
$$\|x'(t)\| \leq \|f(x_0)\|e^{-At}$$

On integrating, we find

$$\|x(t) - x_0\| \leq \|f(x_0)\|/A$$

so that $x(t) \in S(x_0, r)$ for all t.

Thus the solution of (1) exists for all t. In the same way for

$$0 < t \leq t_1, \quad \|x(t_1) - x(t)\| \leq \|f(x_0)\|A^{-1}e^{-At},$$

so that for any sequence $t_n \to \infty$, $x(t_n)$ is a Cauchy sequence and consequently $x = \lim_{t_n \to \infty} x(t_n)$ exists in $S(x_0, r)$.

Moreover, x_∞ is independent of the sequence t_n chosen since clearly

(*) $$\|x(t) - x_\infty\| \le \|f(x_0)\| A^{-1} e^{-At}$$

Also, (6) implies that

$$\|f(x(t))\| = \|x'(t)\| \to 0,$$

and hence $f(x_\infty) = 0$.

The uniqueness of x_∞ follows from this last inequality (*) since if f vanishes at $x, y \in S(x_0, r)$, and $x(t)$ is the line joining x and y, by (3)

$$0 = (f(x) - f(y), x - y) = \int_0^1 (f'(x(t))(x - y), x - y) \ge A\|x - y\|^2.$$

The fact that $F(x_\infty) = \min_{S(x_0, r)} F(x)$ is unique follows similarly since for $x \in S(x_0, r)$, $f(x_\infty) = 0$ and

$$F(x) - F(x_\infty) = \int_0^1 (f(x_\infty + s(x - x_\infty)), x - x_\infty) ds$$

$$= \int_0^1 (f(x_\infty + s(x - x_\infty)) - f(x_\infty), x - x_\infty) ds$$

$$\ge \frac{1}{2} A\|x - x_\infty\|^2.$$

Remark   This result will be extended to hypersurfaces in the Hilbert space H in Chapter 3 on global problems in differential geometry. It will be used there to yield an interesting approach to the Yamabe problem of finding constant scalar curvature Riemannian metrics on compact higher dimensional manifolds.

z = F(x)

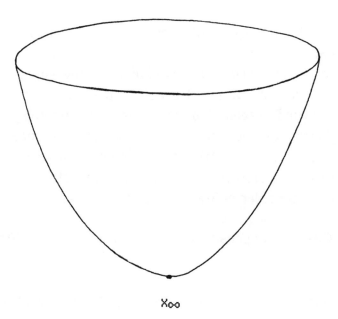

x_∞

Section 2.3   Nonlinear Fredholm Alternatives

Here we consider the operator equation

(1) $Au = f$

Here

(1) A is a C' mapping of a space X into a space Y .
(2) f is a given fixed element of the space Y .
(3) The element u ∈ X is the desired solution of the problem
(1) to be found.

This problem represents one of the classic issues of mathematical nonlinear science. In the linear case, one seeks to find necessary and sufficient conditions for the solvability of (1). Indeed, one of the famous results obtained in the last century, has been the so-called Fredholm alternative, relating the solutions of (1) to the kernel of the operator A. In the nonlinear case, the equation (1) as stated is still of fundamental interest, since it is important to extend this linear Fredholm alternative to a nonlinear context. Let us consider below some simple cases.

Simplest Case   Suppose A = L a given n × n matrix of real constants.

Case A   Suppose L is nonsingular, then L is invertible and (1) has the unique solution $u = L^{-1}f$ for every f .

Case B   Suppose L is singular with finite- dimensional kernel denoted ker L = { $u_1, u_2, \ldots u_k$ } , then (1) is solvable if and only if $f \perp u_i$ $\{i = 1, 2, \ldots k\}$ where $u_i \in \ker L^*$ are a basis for $\ker L^*$ having dimension k.

Now the question arises most naturally: How can this alternative be extended to the nonlinear case? The first possibility is rather straightforward:

Suppose A is a global homeomorphism that is A is $(1 - 1)$ and onto and a mapping from X to Y and that A^{-1} is continuous. Then the solution of (1) exists, is unique and is given by the formula $u = A^{-1}f$.

Here are some criteria that guarantee that A is a global homeomorphism. The main idea of these results is to attempt to extend local results found by analytical means to a global context by the use of topological arguments.

(A) Banach's criteria Suppose (i) A is a local homeomorphism $X \to Y$ and (ii) the mapping A is proper from X onto Y. Then A is a global homeomorphism.

(B) Hadamard's criteria Suppose (i) A is locally invertible and (ii) $\| A'^{-1}(x) \| \leq M$, for all $x \in X$. Then A is a global homeomorphism.

Now we address the more subtle case in which we suppose A is not a global homeomorphism $X \to Y$. In fact we shall suppose that A possesses singular points, i.e. points $\bar{x}$ at which the linearization (i.e. Frechet derivative) $A'(\bar{x})$ is not surjective. To proceed first consider C' nonlinear operators such that the linearization $A'(x)$ always satisfy the linear Fredholm alternative and secondly we distinguish two kinds of points of X on this basis:

I. Regular points x_0 at which $A'(x_0)$ is surjective.
II. Singular points $\bar{x}$ at which $A'(\bar{x})$ is not surjective.

Here is a simple example that has been mentioned in Chapter 1:

Example   (The Periodic Riccati equation) $y' + y^2 = f(t)$ where $f(t)$ is 2π periodic and we seek 2π periodic solutions. Define the Riccati operator $R(y) = y' + y^2$ as a C' mapping between the spaces $W_{1,2}(0,2\pi) \to L_2(0,2\pi)$ with $y(0) = y(2\pi)$ periodic boundary conditions. (see the discussion in Chapter 1 and the articles of the Bibliography)

The Frechet derivative $R'(x)$ is defined by setting

$$R'(x)y = \dot{y} + 2xy \qquad y(0) = y(2\pi)$$

To proceed, we note that the linear operator $R'(x)$ so defined is a Fredholm operator of index zero mapping $W_{1,2}^{(P)}(0, 2\pi) \to L_2(0, 2\pi)$ simply because $R'(x)$ can be written as the sum of an invertible mapping and a compact operator. (This last point follows from the Sobolev compactness theorem). Now to find the regular points and singular point of R (the periodic Riccati operator) we simply look at

$$\dim \operatorname{co} \ker R'(x) = \operatorname{index} R'(x) + \dim \ker R'(x)$$
$$= \dim \ker R'(x)$$

We ask first exactly when is $\ker R'(x) \neq \{0\}$ and in this way determine the singular points x of R. In the same way determining those points x for which $\ker R'(x) = \{0\}$ will yield the regular points. It turns out from simple considerations of ordinary differential equations that a necessary and sufficient condition for $y' + 2xy = 0$ to have a 2π periodic solutions is exactly $\int_0^{2\pi} x(t) = 0$.

This means

(a) If $\int_0^{2\pi} x(t)dt = 0$ and $x(t) \in W_{1,2}(0,2\pi)$ with $x(0) = x(2\pi)$, then $x(t) \in S(R)$ the singular point of R.

(b) If $\int_0^{2\pi} x(t)dt \neq 0$, then $x(t)$ is a regular point of R, assuming $x(t) \in W_{1,2}(0, 2\pi)$ with periodic boundary conditions.

In this situation, we distinguish between local and global analysis. To analyze the local situation we can study the behavior of the solutions of the equation

$$(*) \qquad\qquad\qquad Au = f$$

near a singular point of A, u_0. As f varies over a small neighborhood of $A(u_0) = f_0$? in Y we find the number of solutions of (v). We shall analyze the situation at the simplest type of a singular point u_0, an infinite dimensional fold. In that case, after a change of coordinates we reduce $A(u)$ in the neighborhood of $A(u_0)$ to a global normal form. We can then conclude the real solutions of (A) near u_0 are either 2 or 0 in numbers and in the exceptional case when $f \in$ singular values of A.

Section 2.4   The Idea of Nonlinear Desingularization

Throughout these lectures we shall study novel types of mathematical structures associated with nonlinear phenomena. The examples we choose to consider involve problems from the theory of fluids, geometry, and specific problems in mathematical physics. In many of these areas "nonlinear eigenvalue problems" arise. These are problems that can be written

(1) $Lu = \lambda Nu$

Here the operator L is some linear operator involving elliptic partial differential operators of an unknown function u. N is some nonlinear term, i.e., it involves higher order dependence on the unknown function u , and λ is a parameter to be determined.

If $N(0) = 0$, there is a trivial family of solutions which we choose to be the family $u(\lambda) \equiv 0$. This means that there is a family of solutions that satisfy the equation (1) for all values of the parameter λ, and we normalize these so they can be written as the zero solution.

Linearization Procedures

It is simplest to find appropriate linearizations that approximate the given solutions and then to show that these approximations give rise to actual solutions of the full nonlinear problem. The first linearization is obtained by considering the Frechet derivative at the origin of the system (1). In symbols the associated equation can be written as the linear eigenvalue problem

(2) $Lv = \lambda N'(0)v$

This is the standard linearization for nonlinear eigenvalue problems. We seek nontrivial solutions for the equation (2).

This is a standard linear eigenvalue problem described in Section 2.2. The solutions of (2) called eigenvalues λ and eigenfunction $u(\lambda)$ for a wide class of operators L are well known and tabulated. For the full system (1) the question becomes, How are these solutions of (2) relevant to the solutions of (1)? This is exactly the bifurcation problem described briefly in Section 2.2.

The solutions of (2) for a regular linear eigenvalue problem can be drawn on a plot of (u, λ) axis as follows

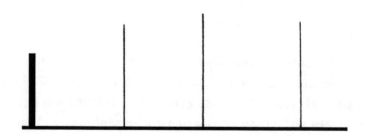

FIGURE 1

The basic idea is that the nontrivial solutions of (2) near $u = 0$ are deformed into solution branches of equation (1). This is illustrated by the following diagram known as a bifurcation diagram

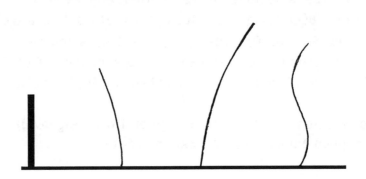

FIGURE 2

This bifurcation diagram illustrates the idea of bifurcation because the solution branch at $u(\lambda) = 0$ bifurcates at λ_1, λ_2, and λ_3 into a pair of solutions that are full solutions of equation (1). The approximate solutions of the nonlinear problem (1) are given by the solutions of equation (2) and they are illustrated in Figure (1). The full solutions of equation (1) are illustrated in Figure (2) above. They show that the trivial solutions $u(\lambda) \equiv 0$ split at λ_1, λ_2, and λ_3 into pairs of solutions with the new solution u_1,λ u_2,λ u_3,λ comprising branches of the solutions for the full nonlinear problem. This process of a solution splitting into two as a parameter changes is known as bifurcation and has become a crucial nonlinear phenomenon associated with scientific problems. It marks from the physical viewpoint a change of stability so that a new family of solutions appears quite suddenly in a given scientific problem as a parameter changes. This means from a realistic point of view that as a scientific problem develops and passes a crucial size, a new and totally unexpected physical appearance often develops. One attempts to understand this mathematically and analytically. This area of nonlinear science is known as bifurcation theory of nonlinear operator equations.

A different linearization procedure can arise in conjunction with equation (1). This process arises by forming a linearization equation as follows

$$(3) \qquad\qquad Lw + c\delta(x) = 0$$

Here $\delta(x)$ is called the Dirac delta function. The equation (3) is an inhomogeneous linear differential equation. It arises from equation (1) as follows. We let the parameter λ tend to an extreme value on the positive real axis, in such a way as the product $\lambda N(u)$ has the following property as λ tends to this extreme value

$$\int \lambda N(u_\lambda) \rightarrow \text{constant } c$$

The equation (3) is considered a linearization of equation (1) in this extreme limit. The solution of (3) is given by the Green's function of the operator L and is a well tabulated and understood quantity. It has the amazing property that it has a singularity at the origin. This means the function G in higher dimensions tends to infinity as $x \rightarrow 0$. In this case the bifurcation diagram does not apply and the linearization is singular in the sense that the solution has infinities in it. Indeed, this singular linearization involves studying solutions of large norm. The effect of the nonlinear equation (1) is to eliminate these singularities and to smoothen the solution u_λ. This phenomenon we call "nonlinear desingularization," in the sense that the singular linearized problem (2) is desingularized by looking at the full nonlinear problem (1). A key question to ask is, What infinities in the approximation can be understood and controlled by looking at the full nonlinear problem in the extreme limit relevant to the problem? Here is a case in point that can be carried out quite explicitly.

Example   Study the positive solutions of the nonlinear Dirichlet problem as $\lambda \to 0$

(*) $\Delta u + \lambda e^u = 0$ on Σ_1

$$u\big|_{\partial\Sigma_1} = 0$$

Here Σ_1 is the unit ball in the plane $\mathcal{R}^2$.

Fact   The system exhibits the nonlinear desingularization phenomena.

Proof   Suppose (as can be proven) the positive solutions u are radially symmetric (i.e. depend only on the radius vector r from the origin). This fact follows from a recent result (Gidas, Ni, and Nirenberg, "Symmetry and related properties via the maximum principle," Communications in Math. Physics vol. 68 , 1979 pp. 209–243). Then we can show that for a real positive parameter β depending on λ, the following explicit solution holds
$$u_\beta(r) = -2 \log (r^2 + \beta)(1 + \beta)^{-1}$$

Thus, after an easy calculation

$$\Delta u = -8\beta(r + \beta)^{-2} = -8\beta(1 + \beta)^2 e^u$$

Thus $8\beta = \lambda(1 + \beta)^2$. Thus we have a quadratic equation for β in terms of λ, and so two roots of this equation.

Notice for fixed positive λ, in the open interval $(0, 2)$ there will be two positive solutions $\beta_+(\lambda)$ and $\beta_-(\lambda)$ for this equation. At $\lambda = 2$, these roots coalesce. In order to display the nonlinear desingularization phenomenon, we study the behavior of the solutions as $\lambda \to 0$. We relate this observation to the explicit solution of the partial differential equation given directly above in the following manner. As $\lambda \to 0$, one branch of solutions, denoted $\beta_+(\lambda) \to \infty$, while the other branch, denoted $\beta_-(\lambda) \to 0$. We observe that the $\beta_+(\lambda)$ root is a reflection of the minimal solutions branch that tends to zero as $\lambda \to 0$. The $\beta_-(\lambda)$ root is a reflection of a second large norm branch of positive solutions of (*). As $\lambda \to 0$ on this "singular" branch we find $\beta_- \to 0$. From the explicit solution of the partial differential equation, we find that as $\lambda \to 0$, $u_{\beta_-}(r)$ converges to $-4 \log r$ (a Green's function).

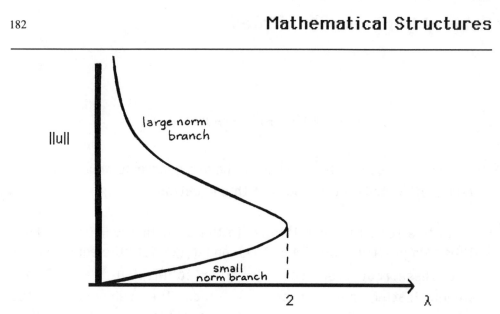

Desingularization Diagram for Positive Solutions of the Nonlinear Dirichlet Problem (*)

From the global point this process can be studied by the calculus of variations. Indeed the positive solutions of (*) on the singular branch can be studied via the isoperimetric characterization.

(π):   Minimize the Dirichlet integral $\displaystyle\int_\Sigma \frac{1}{2}|\nabla u|^2$ over the constraint

$$M_R = \{ u| u \in \dot{W}_{1,2}(\Sigma) \ \text{ with } \int_\Sigma e^u = R \}$$

Here R is a positive parameter.

Note the process of Symmetrization for Calculus of Variations problems of the type (π) is relevant here. By this process, using Steiner symmetrization, we consider a function u_S in M_R that depends only on absolute value of x, i.e. on $|x|$. We know that for u in M_R, u_S produces a function also in M_R, but whose Dirichlet integral is either unchanged or diminished. Thus we conclude that the desired minimizer for (π) is radially symmetric and because of the properties of Sobolev space is actually attained. By the simple regularity theory for the Laplace operator, this critical point is also smooth. Hence, we have an abstract method for producing the desired solution of (π) independent of the explicit solution written above. We conclude that abstract methods can solve such variational problems as (π) without need for explicit solution. Such cases occur throughout nonlinear science.

To see how the process works, note (*) is the Euler–Lagrange equation for (π) with λ in (x) as a Lagrange multiplier. As $R \to \infty$ in (π), the minimizer $u_R(r)$ converges to the Green's function $-4 \log r$.

A) Another Example of Nonlinear Desingularization

The basic ideas of this case can be illustrated by means of a relatively simple example in one dimension. This example arises in the study of free boundary problems associated with fluid vortices. Here the problem consists in finding simultaneously a function U and a connected open set $A \subset (0, 1)$ whose boundary is undetermined, such that

(1.1a)
$$-U_{xx} = \begin{cases} \lambda \text{ in } A, \\ 0 \text{ in } (0, 1)\backslash\overline{A}, \end{cases}$$

(1.1b)

(1.1c, d)
$$U|_{\partial A} = 0, \quad U_x \text{ continuous on } \partial A$$

(1.1e)
$$U(0) = -c, \quad U(1) = -1-c,$$

where λ and c are given (strictly) positive parameters, and $U_x = dU/dx$. Note that (1.1a) and (1.1c) imply that $U(x) > 0$ in A. We can interpret $y = U(x)$ as the equation of a (static) string subject to an upward loading λ (force/length) that is applied only at points where $U(x) > 0$, the ends of the string being fixed at the points $(0, -c)$ and $(1, -1-c)$.

As it happens, (1.1) can be solved explicitly, but of course this is not the usual case, and we first ask, Is there a linear problem that yields approximations to solutions (U, A) of (1.1a)? The answer is yes, but only for a limiting situation as $\lambda \to \infty$. This will be precisely the nonlinear desingularization phenomenon we have in mind.

Assume that the set $\overline{A}$ tends to a point, say a, as the "forcing" $\lambda \to \infty$; then the relevant linear problem is

$$(1.2) \qquad \begin{cases} -V_{xx} = h\delta(x - a) & \text{in } (0, 1), \\ V(0) = -c, \quad V(1) = -1-c, \end{cases}$$

where δ denotes the Dirac delta function; the constants h and a are still to be determined. The solution is

$$(1.3) \qquad V(x) = \begin{cases} -c + (h - ha - 1)x, & 0 \le x \le a, \\ -1 - c + (ha + 1)(1 - x), & a \le x \le 1 \end{cases}$$

The assumption $\overline{A} \to \{a\}$ means that, in the limit, $\max V(x) = V(a) = 0$, which determines $h(a)$. In this particular case, the value a can then be determined from the condition that it minimizes the elastic energy $\dfrac{1}{2}\displaystyle\int_0^1 V_x^2 \, dx$ of the string. Accordingly,

$$(1.4) \qquad h = \frac{c + a}{a(1 - a)}, \qquad a = \frac{c}{2c + 1}$$

Equation (1.3) can be written in a form that extends to other problems:

(1.5) $V(x) = hG(x, a) - q(x)$, $q(x) = c + x$,

where G is the Green's function of the one-dimensional Dirichlet problem for the operator $-d^2/dx^2$, and $-q$ is the solution of $-q_{xx} = 0$ that satisfies the boundary conditions.

How can we find the precise relationship between the solutions of a nonlinear problem like (1.1) and the solution of a linear model like (1.2)? Two conceivable ways are (a) to seek a solution of the full problem by adding small perturbations to the solution of the linear problem, (b) to characterize and bound the solutions of the nonlinear problem globally, and then to analyze their behavior as $\lambda \to \infty$. Here we adopt the second method, which is in accord with our ideas on bifurcation connected with nonlinear eigenvalue problems, we call the "method of nonlinear descent." For generalizations of this problem involving nonlinear partial differential equations, one uses isoperimetric variational principles for functions in a Sobolev space, analogous to our discussion earlier in this section. For the simple problem above, we can relate (1.5) to the explicit solutions of (1.1), which we now describe.

Here is a short explanation and motivation for the term "nonlinear descent." The term "method of nonlinear descent," mentioned above, is a global limiting process to pass from a nonlinear problem to a linearized one. The idea is to try to connect a family of nonlinear problems parametrized by a real number R to a linear problem obtained as the parameter $R \to 0$. For example, in the nonlinear desingularization case studied immediately above we shall introduce a parameter μ below. One is interested in the situation when $\mu \to 0$, since that situation corresponds exactly to the passage from the nonlinear problem (1.1a-e) described above to the linearized problem (1.2) containing the Dirac delta function. In this case, the parameter μ is introduced via the calculus of variations, since this method reduces the study of a saddle point of an infinite dimensional functional to a minimization problem. The nonlinear method is in sharp contrast with power series techniques or perturbation expansions, since these "local" methods attempt to study a nonlinear problem close to a given linear one by expanding out from the linear problem in small convergent arcs to the nearby nonlinear problem.

We now return to the example of nonlinear desingularization just mentioned. Since $U_{xx} < 0$ in A, the set A is indeed connected and consists of a single subinterval of $(0,1)$. Let $A = (a, b)$. First we solve (1.1a-e) separately in the intervals $[0, a)$, (a, b), and $(b, 1]$; then we impose continuity of U and U_x at $x = a, b$. It turns out that, <u>for prescribed</u> λ, there is no solution if $\lambda < \lambda_0 = 8(2c + 1)$, and one solution if $\lambda = \lambda_0$. For $\lambda > \lambda_0$ there are two solutions U_1 and U_2 as in Figure 1, with

$$a_1, a_2 = \frac{c}{2(2c + 1)}\left\{ 1 \pm \left(1 - \frac{\lambda_0}{\lambda}\right)^{\frac{1}{2}}\right\} \text{ respectively,}$$

$$b_1, b_2 = \frac{1}{2(2c + 1)}\left\{ (3c + 1) \mp (c + 1)\left(1 - \frac{\lambda_0}{\lambda}\right)^{\frac{1}{2}}\right\}$$

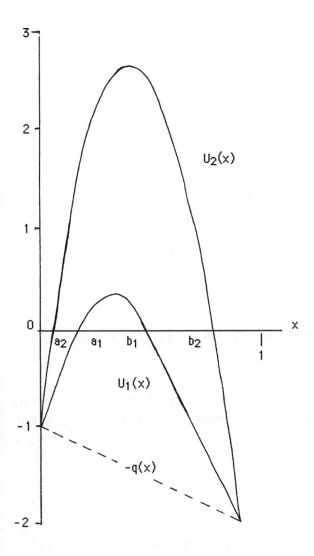

FIGURE 1. The solutions U_1 and U_2 of (1.1) for prescribed λ
$> \lambda_0$. The case drawn is $c = 1$ (for which $\lambda_0 = 24$) and λ/λ_0
$= 1.4$.

Writing

$$U_j(x) = u_j(x) - q(x) \qquad (j = 1,2),$$

where q is as in (1.5) we find that

$$u_j(x) = \begin{cases} \dfrac{c+a_j}{a_j}\, x, & 0 \le x \le a_j, \\[2ex] \dfrac{c+a_j}{a_j}\, x - \dfrac{1}{2}\lambda(x - a_j)^2, & a_j \le x \le b_j, \\[2ex] \dfrac{c+b_j}{1-b_j}(1 - x), & b_j \le x \le 1. \end{cases}$$

From this it is clear that $a_1(\lambda) \to c/(2c + 1)$, $b_1(\lambda) \to c/(2c +1)$, and $U_1(x, \lambda) \to V(x)$ as $\lambda \to \infty$.

    One important point remains. In the variational formulation we shall prescribe, in place of λ, a parameter μ that in the present case is defined by

$$\mu = \int_A U(x)\,dx\, ;$$

then $\lambda(\mu)$ is calculated a posteriori, and nonlinear desingularization occurs as $\mu \to 0$. Here, this has the additional advantage that (1.1) has exactly one solution for each prescribed $\mu \in (0, \infty)$; it is (U_1, a_1, b_1) for $\mu \le \mu_0 = \lambda_0/96$, and (U_2, a_2, b_2) for $\mu > \mu_0$. A calculation shows that $\lambda(\mu)$ is defined (inversely) by

$$\mu = \frac{\lambda}{96} \left\{ 1 - \left(1 - \frac{\lambda_0}{\lambda} \right)^{1/2} \right\}^3 \quad \text{for } \mu \leq \mu_0 \ ,$$

and

$$\mu = \frac{\lambda}{96} \left\{ 1 + \left(1 - \frac{\lambda_0}{\lambda} \right)^{1/2} \right\}^3 \quad \text{for } \mu \geq \mu_0 \ ;$$

the situation is depicted in Figures 2 and 3.

The primary reason for introducing μ, however, is that μ characterizes the set, in a Sobolev space, over which we shall minimize the reduced Dirichlet form of our problem in order to obtain solutions. These are also solutions of an unconstrained variational problem (that is, critical points of a certain functional considered on the whole Sobolev space) but in that formulation, as mentioned above, they are saddle points rather than extrema, making quantitative estimates much more elusive.

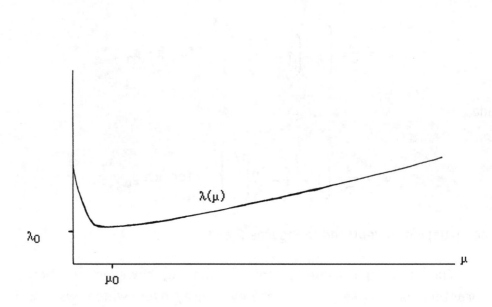

FIGURE 2. Form of the function $\lambda(\mu)$ for the problem (1.1).
The case drawn is
$c = 1 (\lambda_0 = 24, \mu_0 = 1/A)$.

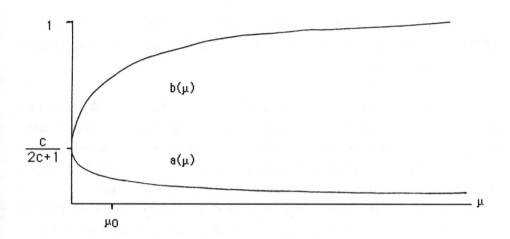

FIGURE 3. Form of the functions a(μ) and b(μ) defining the set A in (1.1). The case drawn is c = 1.

Section 2.5   Variational Principles –
New Ideas in the Calculus of Variations in the Large

We begin this section with the following question already encountered above in this chapter:

(Q): Can the study of "saddle points" (i.e. non-absolute minima) of functionals for problems in geometry and science be ascertained by purely analytical means?

The importance of this problem has increased in recent times. First, the computation of solutions using modern high speed computers has made effective solutions of nonlinear problems a very important part of mathematical analysis of global problems, in spite of the fact that almost all contemporary studies of saddle points are purely existential. Secondly, in order to study the stability of solutions of nonlinear problems, it is extremely important to have as definite as possible a characterization of the critical points of functionals that are under investigation. Thus, a purely abstract existence theorem without explicit analytic computational implementation does not generally enable one to study stability questions for the solution at hand, or to study the properties of the particular solution under investigation. Thirdly, the attempt to sharpen the study of saddle points of functionals by analytical means leads (as we shall see in the sequel) to new methods of analysis that are often deeper and more far reaching than abstract approaches to the same problem. Fourthly, penetrating deeper into explicit analytic means for studying saddle points of functionals often leads to unexpected connections between various aspects of analysis, topology, and geometry. Thus although a solution for a nonlinear problem may be ascertained by minimizing in homotopy classes, that same critical point might be determined by minimizing a functional over a class of functions satisfying explicit constraints that can sometimes be represented as conserved quantities in the relevant physical problem. A fifth point involves the smoothness of saddle point solutions of functionals. It has now become clear due to the researches of Bombieri and Giaquinta, among others, that the regularity (i.e. smoothness) of critical points is more easily determined if the solutions can be represented as absolute minima subject to constraints. In fact, in the sequel, we shall put forward a theory that attempts to represent saddle points of functionals in just these terms, thus obtaining much sharper regularity results for certain problems we address.

It is interesting to consider the recent history of the study of saddle points in light of the question (Q) mentioned above. Most theories of critical points to date are based on their topological determination. The theory due to Marston Morse developed in the last sixty years is based on the use of the Morse inequalities, Morse type numbers, and homology theory, in an infinite dimensional context, for ascertaining the existence of saddle points of broad classes of functionals. The theory of Ljusternik-Schnirelmann is also based on using topological invariants of a type of a cohomology such as category, genus, etc. to establish minimax principles for the existence of saddle points. Both theories are purely existential in nature and have not been used, to date, for constructive purposes. Recently, there are many researches on the so-called Mountain Pass Lemma, a simple minimax principle procedure of the Ljusternik-Schnirelmann type.

In order to find a constructive approach to saddle points, it is helpful to consider a classical precedent. In the linear case, for example, consider the computation of the eigenvalues of the Laplace operator on a bounded domain Ω. There are two points of view for the determination of these numbers. (We note of course, that the critical points of the assumed functionals are simple types of saddle points). The first approach is representing the eigenvalues by appropriate minimax characterizations of saddle points of the Dirichlet integral restricted to a unit sphere in Hilbert space. The second point of view is based on the idea of orthogonal complement, namely, one minimizes the Dirichlet integral iteratively over sets M^N of decreasing dimensions obtained by an orthogonal complement process. Here we consider this last point of view as an example of an answer to the question (Q) raised above. In particular, we inquire about the possibility of generalizing the idea of orthogonal complement to a nonlinear context in relation to the study of critical points of a non-quadratic functional. This idea we call "natural constraints." A description of it now follows.

A Simple Hilbert Space Point of View

The method suggested here, for reducing the study of a saddle point of a functional to a minimization problem with constraints, I call the method of "natural constraints." There, a Hilbert manifold M is taken to be a hypersurface defined by the level sets of one, or more generally, several, appropriately chosen functionals.

By a natural constraint N for a functional I over an admissible class C, I mean a level set of real valued function (or functions) with the following two properties:

(i) Every smooth critical point of I over C satisfies N.
(ii) Every smooth critical point of I over C *and* N, automatically is a critical point over C *unrestricted* by N.

A general construction is given for determining the Hilbert manifold M for a given variational problem. This construction can be used for the example of global families of periodic solutions obtained by the calculus of variations in Chapter 1. In Chapter 3 the idea will be used again in the study of closed geodesics on an ovaloid. However, there, the Hilbert space context is extended to geometric measure theory.

Since the actual construction for natural constraints in the Hilbert space H is straightforward when it applies, I give it here with proof:

Theorem   Let N be a closed linear subspace of a Hilbert space H, and suppose $I(u)$ is a C^2 functional defined on H. Set

$$S = \{u | u \in H, \; I'(u) \perp N\}$$

and suppose

(a) S is closed with respect to weak sequential convergence and nonvacuous,

(b) $I(u)$ is coercive on S and weakly lower semi-continuous there,

(c) $I''(u)$ is definite on N for each $u \in S$.

Then   (i) $c = \inf_S I(u)$ is finite and attained by an element $\bar{u} \in S$, and

(ii) S is a natural constraint for the functional $I(u)$ (on H).

Proof   (i) follows immediately from hypotheses (a), (b), and standard results on minimization. Consequently, we need only establish (ii). To this end, we observe that the elements of S are the zeros of the operator $PI'(u)$ for $u \in H$ (by virtue of the Lax-Milgram theorem). Therefore standard results imply that for an extremal $\bar{u}$ of $I(u)$ restricted to S, there is a fixed element $w \in N$, such that $\bar{u}$ is a critical point of

$$I(u) - (P\,I'(u), w) \quad \text{defined on } H.$$

Here P is the projection operator in Hilbert space from H onto N. Hence the critical point $\bar{u}$ satisfies

$$I'(\bar{u}) = I''(\bar{u})w$$
$$\text{for a fixed element } w \text{ of } N.$$

Taking the inner product of this equation with w, we find that, by virtue of the definition of S,

$$(I''(\bar{u})w, w) = 0, \text{ so } w = 0, \quad \text{by virtue of hypothesis (c)}$$

Thus to verify the fact that S is a natural constraint for $I(u)$ on H, we need only show that every critical point of $I(u)$ (considered as defined on H) lies on S. However, this last fact is immediate since all critical points of $I(u)$ satisfy $I'(u) = 0$ and so $I'(u)$ is necessarily orthogonal to N.

Comments on Utilizing the Theorem

To obtain the natural constraint in the example of Chapter 1.7a on global families of periodic solutions, we merely choose the subspace N to be the kernel of the operator $\underset{\sim}{x}$ with periodic boundary conditions, i.e. the space of constant N-dimensional vectors

$$H = W_{1,2}[(0, 2\pi), \mathfrak{R}^N]$$

For $w \in N$, the natural constraint for this problem becomes

$$(I'(u), w) = \int_0^{2\pi} \nabla V(u(s))ds = 0$$

Similarly, in Chapter 3, we shall find a number of occasions where the notion of natural constraint is quite valuable. For the geometrical applications there involving manifolds, the topological structure of the manifold is often unchanged by the analytic operations involved for a given variational problem. Thus the topological structure can be considered a conserved quantity. When this fact is expressed analytically, we shall find the natural constraint concept is quite useful in reducing critical points of saddle point type to problems concerning absolute minima with constraints.

Here is a case in point which will recur in Chapter 3.

Poincaré's Isoperimetric Problem for Closed Geodesics
(see Berger- Bombieri, Bombieri, and Poincaré in the Bibliography).

Here I wish to discuss one or two the key ideas in our paper with Enrico Bombieri that lead to the solution of Poincaré's isoperimetric problem discussed below and at the same time have the potential for being applied in numerous other contexts in higher dimensions.

Poincaré's Isoperimetric Problem

Poincaré's original geometric problem, in his famous paper of 1905, was to find the shortest nontrivial closed geodesic on a two-dimensional ovaloid M by analytical means. Such geodesics are not absolute minima of the arc length functional defined on M because each point of M can be considered a trivial closed geodesic on M of length zero. Thus the shortest nontrivial closed geodesic on M is necessarily a saddle point of the arc length functional.

Poincaré's isoperimetric problem to find this nontrivial closed geodesic on M of shortest length was the following:

Minimize the arc length of closed curves C on a two-dimensional smooth ovaloid M with Riemannian metric g (i.e. the boundary of a convex body in $\mathfrak{R}^3$), over all simple smooth curves C that bisect total curvature of M and utilize the extremal to determine the shortest nonconstant closed geodesic on the ovaloid.

Before proceeding with a discussion of Poincaré's problem we wish to compare Poincaré's approach with the two other visual approaches to closed geodesics:

(a) **Standard Variational Problem**   Minimize the arc length functional in a homotopy class of smooth closed curves on (M, g).

(b) **Modified Variational Problem**   Utilize the critical point theories of Morse and Ljusternik-Schnirelmann (via Hilbert manifolds) to find the saddle point of the arc length functional that corresponds to the closed geodesic as described above.

(c) **Poincaré's Approach**

Comments on the Various Approaches

(i)   The approach (a) leads to the wrong answer for the ovaloid (viz. the trivial solutions the point geodesics $x(t) \equiv$ point because an ovaloid has $\pi_1(V) = \{0\}$. Indeed, an ovaloid V is simply connected. This is the typical situation of a variational problem with multiple critical points in which the absolute minima of the functional leads to so-called trivial solutions.   It is a good motivation for the need to study saddle points of functionals.

(ii)   The approach (b) has other difficulties: in higher dimensional generalizations since it uses arc length parametrization of curves, it requires special arguments for eliminating the possibility of self-intersections, the approach is basically not analytically effective so that computational and stability questions are not natural in this context. This situation is typical of saddle points determined by purely existential means. The virtue of the method (c) discussed here, and other approaches using natural constraints is that they are closely related to the critical point being studied and can be used to determine more accurate properties of the saddle point under investigation.

(iii)   In the approach (c) differential geometry, in particular the Gauss-Bonnet theorem, instead of topology is used to introduce a strict analytic approach to the problem that enables one to study more specialized properties of the critical point in question.   In particular, its self-intersection properties, its smoothness, and its stability with respect to perturbations in the metric. Moreover, as mentioned in the introduction, this approach requires moving outside the usual Hilbert space of curves to geometric measure theoretic arguments, since point sets $\Sigma_1(C)$ and their boundaries on a Riemannian manifold are the essential variables.

Discussion of Viewpoint (c)

In the language discussed here, the constraint curves $\Sigma_1(C)$ "bisect the total curvature of (M, g)" can be expressed analytically in terms of the Gauss curvature $K(x)$ of the compact 2-manifold (M, g) as follows:

$$\int_{\Sigma_1(C)} K(x) = 2\pi \qquad \text{since the Euler-Poincaré characteristic}$$

$$\text{of } (M, g) = 4\pi$$

This displayed constraint is a "natural" one for the associated variational problem. This fact means

(i) all the smooth simple closed geodesic desired have this property. Here "simple" means that the curves C have no self-intersections.

(ii) adding the natural constraint to the problem of minimizing arc length of curves does not affect the fact that its smooth solutions are geodesics.

Note, however, that this example does not fit in the theory described above because the analysis requires us to go beyond a Hilbert space context.

Proof of (i)   Apply the Gauss-Bonnet theorem to the curve C as in the diagram to find

$$(*) \qquad \int_{\Sigma_1(C)} K + \int_C Kg = 2\pi$$

Now, since C is a closed smooth geodesic on M, the geodesic curvature k_g vanishes identically. Thus the constraint (*) reduces exactly to the words "the curves C bisect the total curvature of (M, g).

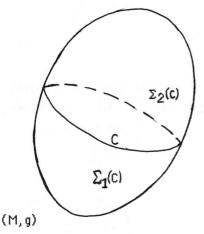

Then since C is a geodesic $Kg \equiv 0$ on C so the natural constraint is satisfied. (Here Kg is the geodesic curvature of C on (V, g)).

Proof of (ii) The Euler-Lagrange equation for the new isoperimetric problem is

$$(**) \qquad Kg = \lambda K$$

where λ is the Lagrange multiplier. Integrating (**) over C we find

$$\int_C Kg = \lambda \int_C K$$

Applying (*) with $Kg \equiv 0$ and the fact that C satisfies the constraint we find

$$\lambda \int_C K = 0 \quad \text{which implies} \quad \lambda = 0$$

since $K > 0$ on an ovaloid.

The topic of natural constraints will arise in very interesting ways in each of the remaining chapters of this book, and we postpone further discussion of this topic to these sections.

Section 2.6   Bifurcation

Suppose that the operator equation $F(u, \lambda) = 0$ defined on a Banach space X, defined over the real numbers, with the parameter λ restricted to be a real number, possesses at least two distinct families of solutions $u_0(\lambda)$ and $u_1(\lambda)$ depending smoothly on a parameter λ, such that $u_1(\lambda) \to u_0(\lambda_0)$ as $\lambda \to \lambda_0$, then, loosely speaking, the pair $(u_0(\lambda_0), \lambda_0)$ is called a bifurcation point of the equation $F(u, \lambda) = 0$ with respect to the primary branch of solutions $u_0(\lambda)$.

More formally, we give the following

Definition   A point (u_0, λ_0) is called a point of bifurcation relative to the equation $F(u, \lambda) = 0$ if (1) (u_0, λ_0) lies on a curve of solutions $(u_0(\varepsilon), \lambda_0(\varepsilon))$ of this equation;   and (2) every neighborhood in $X \times R$ of (u_0, λ_0) has a solution of the equation $F(u, \lambda) = 0$ distinct from this curve of solutions.

In case (u_0, λ_0) is a point of bifurcation for the equation $F(u, \lambda) = 0$, the solutions of $F(u, \lambda) = 0$ near the point of bifurcation often consist of distinct continuous curves through (u_0, λ_0) and these curves are called "branches" of solutions.

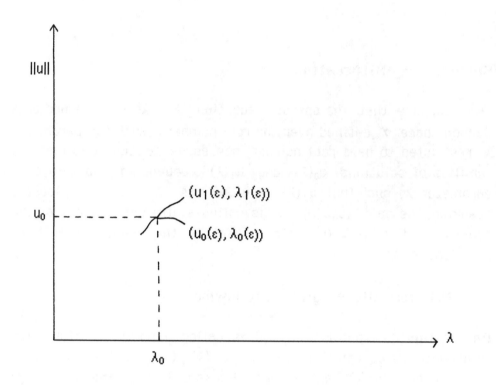

FIGURE. λ_0 is a bifurcation point.

We wish to investigate the relationship between the local behavior of solutions of $F(u, \lambda) = 0$ near $u_0(\lambda)$ and its "linearization" about $u_0(\lambda)$. (By the linearization of an operator $F(u)$ at a point $u_0 \in H$ we mean the operator

$$f'(u_0)u = \lim_{\varepsilon \to \infty}[F(u_0 + \varepsilon u) - F(u_0)]/\varepsilon$$

if this limit exists and is linear in u for all $u \in H$).

Without loss of generality we may assume $u_0(\lambda) \equiv 0$ for all λ. (Otherwise we consider $v = u - u_0(\lambda)$). To this end, we introduce the following restrictions on $F(u, \lambda) = N_1 u + N_2(u, \lambda)$.

(i) N_1 is the identity operator on H (after possibly introducing an equivalent inner product in H).

(ii) N_2 is a completely continuous mapping in u.

(iii) N_2 has a nondegenerate linearization, more precisely, $N_2(u, \lambda) = \lambda(Lu + Nu)$ where L is a bounded linear mapping of H into H and $Lu \neq 0$ for $u \neq 0$, N is a bounded mapping of H into H and $\|Nu - Nv\| \leq f(\|u\|, \|v\|)\|u - v\|$ for all u, v with $\|u\|, \|v\|$ sufficiently small and $f(s, t) = O(|s| + |t|)$ is a real-valued continuous function of the real variables s and t.

With these preliminaries completed, we now state

Lemma A   The bifurcation points of $u = \lambda(Lu + Nu)$ with respect to $u_0(\lambda) \equiv 0$ can occur only at eigenvalues of the linearized equation $u = \lambda Lu$.

Proof   Suppose λ_0 is not an eigenvalue of the linear equation $u = \lambda Lu$, then we shall show that for $|\lambda - \lambda_0| \equiv 0$ and $\|u\|$ sufficiently small, $u = \lambda(Lu + Nu)$ has the unique solution $u \equiv 0$. Indeed, as L is a bounded operator $\|Lu\| \leq \|L\| \|u\|$; and as $I - \lambda_0 L$ is invertible, there is an integer $k > 0$ and independent of $u \in H$ such that $\|I - \lambda_0 L)u\| \geq k\|u\|$. Thus

$$\|u - \lambda_0 Lu\| = \|u - \lambda(Lu + Nu) + (\lambda - \lambda_0)Lu + \lambda Nu\|$$

so that

$$\|u - \lambda(Lu + Nu)\| \geq \|(I - \lambda_0 L)u\| - |\lambda - \lambda_0| \|Lu\| - |\lambda| \|Nu\|$$

$$\geq k\|u\|_m - |\lambda - \lambda_0| \, \|L\| \, \|u\| - O(\|u\|^2)$$

$$\geq (k - |\lambda - \lambda_0| \, \|L\| - O(\|u\|)) \, \|u\|$$

Hence the result follows provided $|\lambda - \lambda_0|$ and $\|u\|$ are sufficiently small, so that $k > |\lambda - \lambda_0| \, \|L\| + O(\|u\|)$.

We now wish to focus attention on the behavior of solutions of $u = \lambda(Lu + Nu)$ for λ near the eigenvalues of $u = \lambda Lu$. To this end we shall need the following lemma.

Lemma B   Let λ_n be an eigenvalue of the compact operator L, then $E_n = \{u | Lu = \lambda_n^{-1} u\}$ has finite dimension.

Proof   Suppose the contrary, namely the existence of a sequence $\{u_m\}$ such that

$$Lu_m = \lambda_n^{-1} u_m \quad \text{and} \quad (u_m, u_k) = \delta_{mk}$$

Then by the compactness of L, $\{Lu_m\}$ has a convergent subsequence which we again label Lu_m ; but

$$\lim_{m,q \to \infty} \|Lu_m - Lu_{m+q}\| = \lim_{m,q \to \infty} \lambda_n^{-1} \|u_m - u_{m+q}\| = 2\lambda_n^{-1} \neq 0$$

a contradiction.

We shall be concerned with certain basic questions regarding the equation $u = \lambda(Lu + Nu)$.

1) **Existence Theory**   The determination of which eigenvalues λ_n of $u = \lambda Lu$ are bifurcation points of $u = \lambda(Lu + Nu)$ with respect to $u_0(\lambda)$.

2) **Multiplicity Theory** The determination of the number of branches of solutions $u(\lambda)$ bifurcating from a given point of bifurcation with respect to $u_0(\lambda)$.

3) **Problem of Nonlinear Invariants**   What qualitative features of the operator N play a significant role in answering the above question (1), (2)?

4) **Problems of Linearization**   What information concerning questions (1), (2) can be determined from a complete knowledge of the solutions of the linearized problem $u = \lambda Lu$?

The following result shows the extent to which the infinite dimensional problems considered here can be reduced to the solution of a finite dimensional problem. It is called the <u>Liapunov-Schmidt reduction</u>.

Theorem C   The solutions of small norm of $u = \lambda(Lu + Nu)$ in the neighborhood of λ_n are determined by the solutions of a finite number (r) of nonlinear equations in (r) unknowns, where r is the multiplicity of the eigenvalue λ_n. These equations are called the bifurcation equations for the problem.

Proof   For simplicity, we assume L is self-adjoint; that is $(Lu, v) = (u, Lv)$ for all $u, v \in H$. In case L is not self-adjoint an analogous but less explicit proof can be given. Choose r linearly independent eigenfunctions of $u = \lambda_n Lu$ and suppose $(u_i, u_j) = \delta_{ij}$ $(i, j = 1, \ldots, r)$. Let the subspace spanned by $(u_1, \ldots, u_r)$ be denoted $[u_1, \ldots, u_r]$ and its orthogonal complement in H be denoted $[u_1, \ldots, u_r]^\perp$. Furthermore, let $P: H \to [u_1, \ldots, u_r]^\perp$ denote the projector of H onto $[u_1, \ldots, u_r]^\perp$. Set $\mu = 1/\lambda$. Thus the totality of solutions of the equations $\mu u - Lu - Nu = 0$ can be obtained by solving the $(r + 1)$ equations

$$P(\mu u - Lu - Nu) = 0$$

(*) $\qquad (\mu u - Lu - Nu, u_i) = 0 \quad (i = 1, \ldots, r)$

Furthermore, any real solution u of this system can be written

$$u = y + \sum_{i=1}^{r} \varepsilon_i u_i$$

where $y \in [u_1, \ldots, u_r]^\perp$ and ε_i are real numbers to be determined. Substituting this expression into (*) we obtain the following equations for y and ε_i

(**) $\qquad y = (\mu I - L)^{-1} P(N(y + \sum_{i=1}^{r} \varepsilon_i u_i))$

(***) $\qquad \varepsilon_i(\mu - \mu_n) = (N(y + \sum_{i=1}^{r} \varepsilon_i u_i), u_i)$

    To proceed further we must assume that we are only concerned with solutions of small norm, that is, that $\|y\|$ and all the numbers ε_i are small compared to 1. The basic idea is to reduce the system (**) – (***) to the finite dimensional study of r equations (not necessarily linear) in r unknowns by showing that the equation (**) is uniquely solvable if the ε_i are sufficiently small. This is accomplished by

Lemma   Under the above hypotheses, and for fixed μ, the equation (**) is uniquely solvable for y in terms of $(\varepsilon_1, \ldots, \varepsilon_r)$. Furthermore, the following estimate holds for the solution $y = y(\varepsilon_1, \ldots, \varepsilon_r)$

$$\|y\| = O(|\varepsilon|^2)$$

where $|\varepsilon| = |\varepsilon_1| + \ldots + |\varepsilon_r|$.

Proof   The solutions of equation (**) are precisely the fixed points of the mapping

$$T_\varepsilon y = (\mu I - L)^{-1} P(N(y + \Sigma \varepsilon_i u_i))$$

By the contraction mapping principle, mentioned earlier in the chapter, this equation has one and only one fixed point provided that for $|\varepsilon|$ sufficiently small.

(i) $\|T_\varepsilon y - T_\varepsilon \bar{y}\| \leq k(\varepsilon)\|y - \bar{y}\|$ where $K(\varepsilon) < 1$ is a positive constant independent of y and $\bar{y}$, and

(ii) T_ε maps the sphere $\Sigma_\varepsilon = \{y|\ \|y\| \leq |\varepsilon|\}$ into itself. Hence we estimate as follows. Let y and $\bar{y}$ belong to Σ_ε, then

$$\|T_\varepsilon y - T_\varepsilon \bar{y}\| \leq \|(\mu I - L)^{-1}P\|\ \|N(y + \Sigma \varepsilon_i u_i) - N(\bar{y} + \Sigma \varepsilon_i u_i)\|$$

$$\leq Kf(\| + \Sigma \varepsilon_i u_i\|, \|\bar{y} + \Sigma \varepsilon_i u_i\|)\ \|y - \bar{y}\|$$

$$= O(|\varepsilon|)\ \|y - \bar{y}\|$$

Thus (i) follows by choosing $|\varepsilon|$ sufficiently small. It remains to show that $\|T_\varepsilon y\| \le |\varepsilon|)$ for $\|y\| \le |\varepsilon|$. Indeed,

$$\|T_\varepsilon y\| \le Kf(\|y + \Sigma\varepsilon_i u_i\|), 0) \|y + \Sigma\varepsilon_i u_i\|$$

So $\|T_\varepsilon y\| = O(|\varepsilon|)^2$, and $\|T_\varepsilon y\| \le |\varepsilon|$ for $|\varepsilon|$ sufficiently small. Hence T_ε is a contraction mapping defined on the sphere $\|y\| \le |\varepsilon|$. Furthermore, we prove the following estimate for the solution of $T_\varepsilon y = y$ with $\|y\| \le |\varepsilon|$:

$$\|y\| = T_\varepsilon y - T_0 y + T_0 y\|$$

$$\le T_\varepsilon y - T_0 y\| + \|T_0 y\|$$

$$\le K[\|N(u) - N(y)\| + \|N(y)\|]$$

$$\le K[f(\|u\|, \|y\|) \|u - y\| + f(\|y\|, 0) \|y\|]$$

Hence $\|y\| = O(|\varepsilon|)^2$ and this complete the proof.

The final step in the proof of Theorem C is the observation that for given μ the system (**) – (***) can be written in the form

$$\varepsilon_i(\mu - \mu_n) = f_i(\varepsilon_1, \dots, \varepsilon_r) \qquad (i = 1, \dots, r)$$

where the continuous functions f_i are completely determined by $(\varepsilon_1, \dots, \varepsilon_r)$, μ, and the solution y is determined by the lemma.

Remark   If hypothesis (iii) is strengthened by assuming $\|Nu - Nv\| \le K\|u\|^p + \|v\|^p\} \|u - v\|$ for constants K and $p > 1$ and independent of u and v, then the estimate of Lemma is $\|y\| = O(|\varepsilon|^{p+1})$.

The above method is interesting because for many qualitative problems an explicit knowledge of y is unnecessary. For example, we have the following corollary.

Corollary Suppose

(i) $(N,(u), u) > 0$ for $u \neq 0$,

(ii) N is a completely continuous mapping homogeneous of degree $p + 1$ that is $N(\sigma u) = \sigma^{p+1} Nu$ for fixed positive p, and

(iii) the hypothesis of the above remark is satisfied. The any nonzero solution of small norm of $u = \lambda(Lu + Nu)$ with λ in the neighborhood of λ_n occurs with $\lambda < \lambda_n$.

Proof Since $u \neq 0$, $|\varepsilon| \neq 0$; for if $|\varepsilon| = 0$, the estimate of the Lemma above implies $y = 0$ and as all small solutions are of the form $u = y + \Sigma \varepsilon_i u_i$, $u \equiv 0$. By Theorem C, it is sufficient to consider the r equations (***)

$$\varepsilon_i(\mu - \mu_n) = (N(y + \Sigma \varepsilon_i u_i, u_i)$$

Furthermore, these equations can be written

(†) $$\varepsilon_i(\mu - \mu_n) = (N(\Sigma \varepsilon_i u_i, u_i) - (R, u_i) \quad (i = 1, \ldots, r)$$

where $R = N(y + \Sigma \varepsilon_i u_i) - N(\Sigma \varepsilon_i u_i)$. By the Cauchy–Schwarz inequality, hypothesis (iii), and the estimate of the Lemma

$$|(R, u_i)| \leq \|R\| = O(|\varepsilon|^{2p+1})$$

Now multiplying each equation of the system ($'$) by ε_i and summing

$$(\Sigma \varepsilon_i{}^2)(\mu - \mu_n) + (N(\Sigma \varepsilon_i u_i), \Sigma \varepsilon_i u_i) + (R, \Sigma \varepsilon_i u_i)$$

From the above, we find that

$$|(R, \Sigma \varepsilon_i u_i)| = O(|\varepsilon|^{2p+1})$$

and to show that $\mu > \mu_n$, we estimate the first term on the right of the above equation as follows. Define $\varepsilon_i = |\varepsilon|\bar{\varepsilon}_i$ for then $|\bar{\varepsilon}| = 1$ and by hypothesis (ii) $N(\Sigma \varepsilon_i u_i) = N(|\varepsilon| \sum_i \bar{\varepsilon}_i u_i) = |\varepsilon|^{p+1}N(\sum_i \bar{\varepsilon}_i u_i)$. On the other hand, by hypothesis (i)

$$(N(\Sigma \varepsilon_i u_i), \Sigma \varepsilon_i u_i) = |\varepsilon|^{p+2}(N(\Sigma \bar{\varepsilon}_i u_i), \Sigma \bar{\varepsilon}_i u_i)$$

$$\geq |\varepsilon|^{p+2} \inf_{|\bar{\varepsilon}_i|=1} (N(\Sigma \bar{\varepsilon}_i u_i), \Sigma \bar{\varepsilon}_i u_i)$$

$$\geq C |\varepsilon|^{p+2}$$

where $C > 0$. Therefore from the above inequalities

$$(\Sigma \varepsilon_i{}^2)(\mu - \mu_n) \geq C |\varepsilon|^{p+2} - k_0'' |\varepsilon|^{2p+1} = |\varepsilon|^{p+2} [C - k_0'' |\varepsilon|^{p-1}]$$

Thus for sufficiently small $\varepsilon \neq 0$, $(\mu - \mu_n)$ is positive and so for a nonzero solution of the system (**), $\mu > \mu_n$, that is $\lambda < \lambda_n$.

Section 2.7   Bifurcation into Folds

Introduction

It is remarkable how many times the notion of computing the behavior of a nonlinear equation near a simple singular point arises. It is this simple singular point that we call a "fold" in this book. In Chapter 1 this idea arose when we tried to find the simplest nonlinear global normal form that incorporated bifurcation phenomena. What is remarkable about this notion of a fold and bifurcation is that the notion of a fold does not involve a bifurcation parameter in any explicit way. In particular, many aspects of the numerical analysis of nonlinear equations, mathematical physics and applied mathematics relate to this topic. Thus, here we present a collection of examples on how infinite dimensional simple singular points come up in infinite dimensional nonlinear analysis and its applications.

The simplest example is in the classic Riccati equation, as mentioned in the above sections. Here we consider a simple generalization. The next example we introduce here is the following iteration process for real numbers of the form $x_{n+1} = f(x_n)$ as described above. Here we choose

$$x_{n+1} = k + x_n - x_n^2 \qquad \text{where } k \text{ is a constant}$$

There are two important cases

    (i) $k < 0$, in which case the iteration has no fixed points at all
    (ii) $k > 0$ and small, in which case the iteration has two fixed points $\pm \sqrt{k}$. As before the stability criterion is obtained

by setting $|f'(x)| = |1 - 2x|$, so that for small positive k,
the fixed point $\sqrt{k}$ is stable, whereas the fixed point $-\sqrt{k}$
is unstable.

The next example is certain nonlinear Dirichlet problems, and the
final example that we discuss will be in semiconductive device
analysis in conjunction with switching phenomena.

The Periodic Generalized Riccati Equation

We consider the equation

(1) $\qquad\qquad \dot{x} + g(x) = f(t) \qquad$ where $\dot{x} = \dfrac{d}{dt}$

where $g(x)$ is a strictly convex smooth function of x and so
resembles x^2 of the Riccati equation. Here $f(t)$ is a given forcing
term of fixed period, say T. We seek the solutions $x(t)$ that are
also T-periodic. There is a large study of Liouville that goes into
detailed analysis of various functions $f(t)$ for which this equation
cannot be solved by quadrature. These cases can be found in Chapter
4 of Watson's <u>Treatise</u> <u>on</u> <u>Bessel</u> <u>Functions</u>.

This means, as mentioned in Chapter 1, that the Riccati
operator

$$A(x) = \dot{x} + x^2$$

for example, cannot be linearized by appropriate change of
coordinates. It is important to analyze why this is true for the
equation (1). The answer is very simple:

(1)   Every singular point of the periodic Riccati operator is a simple singular point, that is an infinite dimensional fold.

(2)   There is an infinite dimensional hyperplane of singular points for the operator A regarded as a nonlinear mapping between spaces of periodic functions as defined below.

We now consider the generalized Riccati equation defined above in equation (1). It will turn out, that simple extensions of these two facts mentioned above will also hold for the generalized Riccati equation. Here are some facts about the equation (1) and the associated nonlinear operator

$$Bx = \dot{x} + g(x)$$

Our first observation about B is that

1)   B is a bounded nonlinear mapping of the T-periodic functions of the Banach space $W_{1,2}$ into L_2.

Proof   Expanding the L_2 inner product and using Jensen's inequality we find that

$$\|B(x)\|_{L_2} = ((\dot{x} + g(x), \dot{x} + g(x))^{1/2}$$

Consequently,

$$((\dot{x} + g(x), \dot{x} + g(x)) = \|\dot{x}\|_{L_2}^2 + \int_0^T g^2[x(s)]ds$$

Here we have used the fact that the cross product term vanishes, since the associated integrand is a perfect differential. Hence, we find that whenever the norm $\|x\|$ in the Sobolev space $W_{1,2}$ is bounded by a constant M, say, then

$\|B(x)\|_{L_2}$ is uniformly bounded

Here the periodic boundary conditions $x(0) = x(T)$ are used to prove that if $G'(x) = g(x)$,

$$\int_0^T g(x)(s)\dot{x}ds = \int_0^T \frac{d}{ds} G(x(s))ds = 0$$

i.e. as mentioned above, the cross product terms in the above expansion vanish.

Now we prove

2)   B is a nonlinear Fredholm operator with index 0. That is, it's Frechet derivative, $B'(x)$, is a linear Fredholm operator with index 0, at each point $x \in X$.

Proof We denote the Frechet derivative operator

$$B'(x)y(s) = \dot{y} + g'(x)y(s)$$

Clearly $B'(x)$ is a bounded linear operator acting between the Sobolev space $W_{1,2}(0,T) \to L_2(0,T)$ with T-periodic boundary conditions. Now we rewrite $B'(x)$

$$B'(x)y = (\dot{y} + \varepsilon y) + (g'(x) - \varepsilon)y$$

Thus, using the Sobolev-Kondrachev inequality, $B'(x)y$ can written as the sum of an invertible plus a compact linear operator. Hence $B'(x)$ is a Fredholm operator of index zero.

Note that for $\varepsilon > 0$ the linear operator $Ly = \dot{y} + \varepsilon y$ is an invertible operator between the above Hilbert spaces of periodic functions.

Hence the operator B itself is a nonlinear Fredholm operator of index zero.

Now we prove that the dimension of a kernel of $B'(x)$ is at most 1 and dim ker $B'(x) = 1$ exactly when

$$\int_0^T g'(x(t))dt = 0$$

Proof (i) The fact that dim ker $B'(x) \leq 1$ is an immediate consequence of the elementary linear theory of ordinary differential equations.

(ii) To investigate the precise kernel of $B'(x)$ for $x(s) \in W_{1,2}$ we solve the following equation for $y(s)$ with T-periodic boundary conditions

$$y' + g'(x)y(s) = 0$$

where

$$g'(x(0)) = g'(x(T))$$

to obtain

$$y(s) = y_0 \exp\left(\int_0^s g'(x(t))dt\right)$$

We decompose the function g'(x(t)) into its mean value k   and orthogonal part h(x(t)) of mean value zero over (0, T). Thus

$$y(s) = y_0 \exp \{ks + \int_0^s h(x(t))dt\}$$

Moreover, we find, using the fact that $\int_0^T h(x(t))dt = 0,$

$$y(s + T) = y_0 \exp \{ks + kT + \int_T^{s+T} h(x(t))dt\}$$

$$= e^{kT}y(s)$$

Thus y(s) will be T-periodic exactly when $e^{kT} = 1$. Equivalently, we require for T-periodicity of y(s) that $K \equiv 0$, since the other alternative T = 0 is inadmissible. Thus

$$\int_0^T g'(x(s))ds = 0$$

In words, this last equation means that the mean value of g'(x(s)) over a period vanishes.

Let us   prove that each singular point of the periodic generalized Riccati equation is a simple singular point (that is, a Whitney fold).

To this end we   again use the following theorem, as mentioned in Chapter 1, proved by myself and P. Church.

Theorem 1   Let A be a C^2 Fredholm map of index zero between two Hilbert spaces H_1 and H_2 with a singular point at x. Suppose

(i)   dim ker A'(x) = 1 and

(ii) $(A''(x)(e_0,e_0),h^*) \neq 0$, where $e_0 \varepsilon$ ker A'(x) and
       $h^* \varepsilon$ ker[A'(x)]* are both not identically zero. Then A is
       a local Whitney fold near x.

This implies after a change of coordinates that A can be written (near x) as $(t, v) \to (t^2, v)$. *This means that A has a simple singular point at x if the linearized mapping has a one-dimensional kernal and its second derivative is non-degenerate.*

Now, we compute B'(x)y near a singular point x,

$$B'(x)y = dy/dt + g'(x)y$$

with the same T-periodic boundary conditions for y. Thus, as above, B'(x)y can be represented as the sum of invertible linear map plus a compact map. This fact establishes out first goal. Then we classify each singular point of B.   Now by Taylor series expansion

$$B(x + h) = B(x) + [g'(x)h] + \frac{1}{2} g''(x)h^2 + o(\|h\|^2)$$

This shows that the second derivative of B at x, $(B''(x)(h, h) = g''(x)h^2$ is "nonnegative" and strictly positive, by virtue of the strict convexity assumed for g(x) and the fact that $h \neq 0$. Moreover, the kernel of the adjoint of [B'(x)], namely, [B'(x)]* equals the T-periodic solutions of the equation

$$y' - g'(x)y = 0$$

that do not vanish identically are one signed. Thus the required L_2 inner product of

$$(B''(x)(h, h), \ h*)_{L_2} \neq 0$$

Thus Theorem 1 above applies, so every singular point of B is a Whitney fold and so has the desired local normal form.

The remarkable fact about our work on the periodic generalized Riccati equation is that this problem, although not integrable by quadrature, can be studied globally. In fact, the mapping B is a global Whitney fold. This is discussed in the paper of McKean and Scovil mentioned in the Bibliography.

It shows that, in theory at least, the Riccati operator A and the generalized Riccati operator B with periodic boundary conditions can be studied theoretically from a global point of view using the ideas of normal forms even though bifurcation in the form of a simple singular point presents itself many times in the problem. This is a new idea but should be useful in computations. Another important feature relevant to numerical computation work is that the methods used in studying a simple singular point are <u>stable</u>. This view that if the Riccati equation (1) is perturbed slightly by changing the nonlinearity from x^2 to $g(x)$, then the periodic solutions of the problem are not altered significantly, but are merely perturbed. The same theoretic structures apply because they are independent of perturbation. In fact the global normal form remains invariant, only the global changes of coordinates are slightly perturbed.

§2. Nonlinear Dirichlet Equation

We consider again nonlinear partial differential equations on a bounded domain R^n with null boundary conditions imposed on the boundary. This equation can be written

(*) $$\Delta u + f(x,u) = g(x)$$

$$u\big|_{2\Omega} = 0$$

This equation is conditioned by the fact that f has the following asymptotic properties

$$\lambda_2 > \lim_{t\to\infty} \frac{f(x,t)}{t} > \lambda_1 \; ; \; \lambda_1 < \lim_{t\to-\infty} \frac{f(x,t)}{t} > 0$$

where λ_1 and λ_2 are the lowest two eigenvalues of L relative to Ω.

Once again we proved in Chapter 1.4 that if the equation is written Au = g then the associated operator A is a Fredholm operator of index 0 acting between appropriate Sobolev spaces. In fact one can prove an analogous result to Chapter 1. In particular, we show the following facts:

(1) Every singular point of the operator A is a simple
     singular point or infinite dimensional fold
(2) The singular points of A on a hypersurface of
     codimension 1, in the Sobolev space H.
(3) The equation (*) has either 2, 0, or 1 solutions depending
     on whether g is a singular value or not, of the operator
     A

(4)   Theoretically the solution can be written down explicitly
       in terms of compatible quantities for equation (1)

The same global properties that applied to the periodic-Riccati equation apply for the problem discussed here as well, independent of the shape of the domain Ω or the number of spatial dimensions involved. Thus it is quite remarkable and will undoubtedly lead to new developments in the numerical structure of nonlinear partial differential equations. In particular, the associated mapping A is a global Whitney fold; i.e. as the structure of a simple singular point namely it can be converted after changes of coordinates to a simple quadrature mapping.

This idea can also be carried out for nonlinear parabolic equations of the form (see Appendix 2 at the end of the chapter)

$$\frac{\partial u}{\partial t} + \Delta u + f(x, u) = q(t, x)$$

$$u\big|_{2\Omega} = 0$$

$$u(x,t) \quad \text{is T-periodic in t}$$

Here we require that the function $q(t,x)$ be T-periodic in t.

Thyristor: Folds in Semiconductive Devices

The third example of fold structures enters into switching devices for the Thyristor. The semiconductor device equations can be written

$$Au(x) = f$$

where the semiconductor device has an interesting doping profile given by the inhomogeneous term f and A represents a system of nonlinear partial differential equations. It turns out that the J-V curve measuring current-voltage for this semiconductor device can be pictured on the diagram below when the doping f is sufficiently large. Then the device has one equilibria or three depending on the intersections of V with the current axis written above. This gives rise to the switching phenomenon we have in mind, because the doping profile is an S shape curve at exactly the points at which the S shape has minimum curvature on the J-V curve . Then at exactly the points a fold structure occurs. These fold structures can be precisely arranged by an external gate.

More explicitly, the simplest semiconductor device exhibiting this behavior is the thyristor. The thyristor exhibits a switching phenomenon that has been explained in physical terms but whose mathematical development up to now has been unclear. In fact, in terms of a J-V curve which is drawn below, the group of the voltage vs. current exhibits negative resistivity.

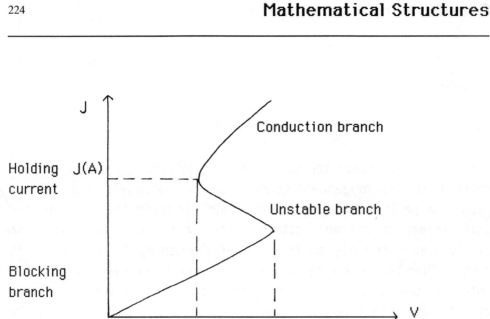

FIGURE 1   J-V curve for the thyristor.

The mathematical explanation of the switching process is as a bifurcation phenomenon. The bifurcation equations are a consequence of the mathematical structures inherent in the equation and in the boundary conditions.   The important parameter involved is the magnitude of the doping profile (see below).

The important issue mathematically is the description of the behavior at the bifurcation point. The nonconvexity of the J-V graph can be made more precise by an analysis of the bifurcation problem posed in mathematical terms.   Each bifurcation point is a fold and this leads to a justification of the turning points A and B on the V-axis.

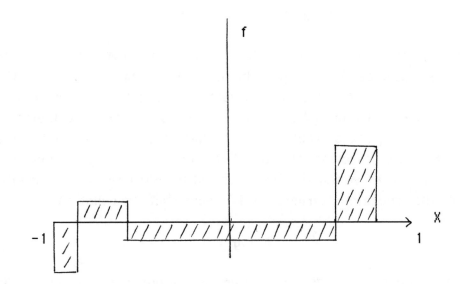

FIGURE 2. Doping profile for the thyristor.

The doping profile generating the J-V curve of Figure 1 is drawn in Figure 2. It consists of dividing the x-axis into four segments and alternating the sign of the doping profile as indicated in the figure. This doping profile coupled with the 1-D steady state van Roosbroeck equations generates the J-V curve of Figure 1.

The Mathematics of Bifurcation Phenomena: For any physical system described by a system of the form $Au(x) = f$, we can regard the differential operators and boundary conditions A as a mapping between two function spaces X and Y. Generally one studies the linearization of A, $A'(u)$ at a function u simply by writing the Taylor series

(2) $A(u+\varepsilon y) = A(u) + \varepsilon A'(u)y + O(|\varepsilon|^2)$

Here y is another function and ε is a real number. In finite
dimensional case $A'(u)$ is just the Jacobian matrix evaluated at the
vector u and A is simply a vector of functions. For partial
differential operators, $A'(u)$ is the linearized operator evaluated at
u. For steady state problems as in the form (2) describing the
thyristor we describe the onset of bifurcation by finding the
functions u at which the linear operator $A'(u)$ is not invertible.
Using the Fredholm alternative this means that the equation

$$A'(u)y = 0$$

supplemented by boundary conditions has a nonzero solution y. If
the equation $A(v) = f$ has non-unique solutions for v near u and
f near $A(u)$ we call u a bifurcation point.

If $A'(u)$ is always invertible no bifurcation exists, and for
computation purposes the Newton method or a modification of it
suffices for computing numerical results.

If bifurcation does occur at u, i.e. u is a bifurcation point,
there is a classification of different kinds of behavior based on the
structure of the differential operator A. The simplest case ,as
described here and in the previous sections, is called a fold and can
be characterized by the fact that near the bifurcation point u (say)
all linear approximations to the behavior of A fail, but (after a
change of coordinates) A can be rewritten near u in the form (t,
y) → (t^2, y). This means that after a change of coordinates A has a
parabolic shape.

In terms of the operator A the fold is easily distinguished by virtue of Theorem A, described above.

The switching behavior of the thyristor can be easily explained via the J-V curve. When the voltage V, the biasing, is switched on and is small, the thyristor has small J as shown in the "blocking branch" before A in Figure 1. To get the thyristor into the conduction branch with high J-current without altering V, one utilizes the fact that the "middle branch" joining the blocking and conduction branch is unstable. Thus introducing a sufficiently large perturbative gate current to the system in the blocking state causes the system to jump "over" the unstable branch into the conducting state. Conversely, once in the conduction state the system can jump to the blocking state by reducing V below the holding current voltage A (as in Figure 1).

Chapter 3   Some Connections between Global Differential Geometry and Nonlinear Analysis

For many centuries the areas of geometry and analysis were considered as separate entities. With the advent of far-reaching results in linear analysis, very important linear structures were found throughout geometry. These structures were linked with questions in both geometry and analysis. This link between the two fields can be seen in the theories of Hodge and the spectral theory of operators, e.g. the Laplace operator on manifolds. When a more global point of view is taken, the linear structures inherent in geometry problems can sometimes be supplemented by adding nonlinear effects. The key nonlinear effect inherent in geometry is the notion of curvature in all its ramifications. This notion is a fundamental source of deep problems involving global problems of geometry and fundamental aspects of nonlinear analysis.

Here in this chapter we shall discuss the beginning aspects of this connection and look at basic issues involving Riemannian and Kähler metrics on manifolds as well as special curves on manifolds such as closed geodesics. As another simple example, we consider a compact two-dimensional manifold such as one observes in everyday life such as a sphere or torus T^2. We ask the very simple but basic question: What is the simplest metric structure that can be defined on this manifold?Can this metric structure be characterized by an extremal principle?

It should be stated that this procedure is in a certain sense an "inverse" problem. Normally one defines an extremal principle, and then finds critical points of this principle which automatically satisfy the associated variational equations, the so-called Euler-Lagrange equations. Here we take a reverse viewpoint. One assumes as given the equations that one wishes to satisfy. One hopes that there is a functional F with a property that these equations are the Euler-Lagrange equations of this functional. Then one attempts to solve the equations not directly but by the simple method of finding critical points for the functional F. In order to carry through this procedure one must of course require that the equations in some sense be associated with the smooth critical points of some unknown functional F. As mentioned in Chapter 2, in terms of functional analysis, such equations are called gradient operator equations, because of the simple fact that they must be represented in some general sense by an operator equation in a Hilbert space that is derivable from a real valued functional.As mentioned at the end of Chapter 1, the Solution Diamond Diagram summarizes some of the main ideas here. Consider the following simple examples:

(A) $\text{grad } F(u) = g$ where g represents a given inhomogeneous term, found from given geometric data

(B) $\text{grad } F(u) = \lambda \text{grad } G(u)$ where λ is a constant (a nonlinear eigenvalue)

The variational problems associated with (A) and (B) can be easily stated. For (A) the problem involves finding the critical points of the functional $F(u) - (g, u)$ defined on a Hilbert space H. On the other hand, for (B), λ is regarded as a Lagrange multiplier and we search for critical points of the functional $F(u)$ restricted to the infinite-dimensional manifold M, an infinite dimensional hypersurface on H defined by the equation $G(u) = $ a constant. All these interpretations occur, for example, when we suppose the element u belongs to a Hilbert space H or some more general function space.

Of course, the great virtue of geometry is that generally these functionals have a geometric connotation. This is sometimes the case, but we shall find instances when we are necessarily forced to use methods of abstract analysis to find the associated geometric functional F. Moreover, the function spaces associated with the calculus of variations approach outlined in this book, need to be chosen with some art. This marks one of the advances in contemporary analysis. In fact, appropriate function spaces for the geometric problems discussed in this book will be fairly evident. The intimate connection between the function spaces X and the associated functional F comprises an area of investigation with many outstanding open problems.

A key difficulty, known to many researchers, in utilizing the global ideas I have in mind, is that, for example, for the case of closed geodesics, interesting periodic motions are not usually absolute minimizers of the Lagrangian of nonlinear Hamiltonian systems, but rather appear as saddle points in either the Lagrangian or Hamiltonian formulations. The mathematical study of periodic motions by the global calculus of variations ideas described above, is then forced to consider mathematical theories concerned with the so-called critical points of functionals, an area of mathematics that is distinguished by the combined use of all the resources of modern research, namely a combination of analysis, algebra, and topology. There are basically three theories of critical points of functionals: (I) Morse theory, (II) Ljusternik-Schnirelmann theory, (III) the method of Natural Constraints. The first two areas are described in many books mentioned in the bibliography, the third area is discussed in Chapter 2.

Section 3.1   Geodesics

I wish to begin with an example of the last method of natural constraints from a historical point of view dating back to Poincaré's research ideas of 1905. This problem illustrates the fact that sometimes new analysis needs to be developed from a very clear-cut geometric situation.

Problem 1   A One-Dimensional Problem

Problem   (Poincaré 1905):   Find   the   shortest   closed (nonconstant) geodesic on a smooth ovaloid.   Determine its properties.

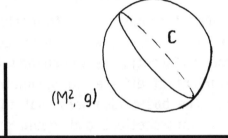

Figure A   A Typical Closed Geodesic on an Ovaloid M

This problem is connected with periodic motions of conservative systems because we are searching for nonconstant closed geodesics. Indeed, a closed curve is generally represented analytically by periodic functions. The manifold (M^2, g), a two-dimensional manifold with Riemannian metric g in our case, is chosen to be an ovaloid that is the boundary of a convex bounded two-dimensional set, and one can see that the natural problem of minimizing the arc length functional on closed curves yields zero as minimum.   Zero corresponds to the noninteresting closed geodesics, the solution in which the curve C reduces to a point. Thus the interesting periodic solution is thus not an absolute minimum of the arc length functional. In fact, it is a saddle point of this functional, and to determine this geodesic in a <u>constructive</u> way has proved to be, in the nearly eighty years since Poincaré's approach, a difficult problem.

The natural global topological invariant for the problem, namely the homotopy class of curved curves on M^2, is zero because M^2 is an ovaloid, and so possesses the same homotopy as the two-sphere.   Thus since $\pi_1(M^2) = 0$,   a substitute must be found for homotopy. Moreover, this substitute must be sufficiently strong to reveal the properties of the shortest nonconstant closed geodesic C. Poincaré's idea was to use differential geometry and to replace homotopy theory by an invariant of differential geometry, namely the Gauss-Bonnet theorem.   The basic idea was that this special geodesic would have no self-intersections and would possess (in restricted cases) special stability properties as a geometric object.

To begin with a mathematical treatment, we assume a simple closed curve C exists, divides the manifold into two simply connected pieces Σ_1 and M $-\Sigma_1$ and apply the Gauss-Bonnet theorem to it (see Figure A). We obtain

(1)
$$\int_{\Sigma_1} K + \int_C K_g = 2\pi$$

Here K is the Gauss curvature and K_g is the geodesic curvature. Now we assume that C is a geodesic so that $K_g = 0$ and we obtain from (1)

$$(2) \qquad \int_{\Sigma_1} K(x)dV = 2\pi$$

This equation is the invariant we seek. In fact, Poincaré chose it. It is the natural constraint we are searching for because (as required by the theory of Chapter 2) we can prove that minimizing the arc length functional relative to the constraint (2) yields a simple closed geodesic, and on the other hand, as we have shown above, every simple closed geodesic on (M^2, g) satisfies (2). This complete analysis was first carried out a few years ago in my paper with E. Bombieri, mentioned in the Bibliography, for ovaloids with curvature restrictions, and has subsequently been shown to be valid without such restrictions by Allard.

For example, here is the proof that shows that a smooth minimum of the arc length functional constrained by (2) is a closed geodesic. The Euler-Lagrange equation for a smooth critical point of the arc length functional subject to (2) is

$$(3) \qquad K_g = \lambda K \quad \text{where } \lambda \text{ is a Lagrange multiplier.}$$

Integrating this equation over the smooth curve C we find

$$(^\wedge) \qquad \int_C K_g = \lambda \int_C K$$

Utilizing the Gauss-Bonnet theorem (1)

$$\int_{\Sigma_1} K \; -2\pi \; = \; \lambda \int_C K$$

Utilizing the constraint (2), we find, from the above equation, that

$$\lambda \int_C K = 0$$

Since (M^2, g) is an ovaloid, $K > 0$ on C, so this last equation implies that $\lambda = 0$.

Thus the Euler-Lagrange equation (3) reduces to

$$K_g = 0$$

The vanishing identically of the geodesic curvature K_g implies by standard differential geometry that C is a geodesic.

This result implies that the constraint (2) does not affect the Euler-Lagrange equation for geodesics and it is therefore (in our terminology) called a natural constraint. Thus the shortest nonconstant closed geodesic can be characterized as a minimum of the arc length functional over a general class of closed curves subject to this natural constraint.

This idea of using isoperimetric variational principles has a number of virtues in this special case. First, the minimizing curves in question can be shown to be smooth and have no self-intersections by virtue of their minimizing properties. Secondly, such extremal curves can be computed by constrained optimization methods. Thirdly, the stability of such geodesics can be studied carefully by perturbing the metric g. Fourth, this idea can be extended to nonlinear dynamical systems and periodic solutions. In fact, in Chapter 1 we showed how to extend this idea to other nonlinear Hamiltonian systems to find large amplitude periodic solutions.

The procedure that Poincaré suggested in his study of this problem thus consisted of four steps:

1) Find the extremal for the isoperimetric problem. (II) Minimize the arclength among all smooth closed curves {C} satisfying (2). Prove the extremal is attained in some generalized class of curves.

2) Prove the resulting extremal is smooth.

3) Prove the resulting extremal is non self-intersecting.

4) Study the stability of this extremal, regarded possibly as a simple periodic orbit of a dynamical system.

Poincaré at the beginning of the twentieth century was unable to carry out the details of this program, because the analysis, needed for this work, had not been developed. But currently, the details can be carried through in an effective analytical approach. What is interesting is that the ideas involved require a special twist. They have been developed but are relatively inaccessible in current mathematical literature. Thus it is important to view the details as well as the spirit of the new analytic notions involved. The basic idea is the utilization of the notion of integral current. It is in terms of integral currents and their generalizations that allow the details of this problem of Poincaré's to be carried through.

We review the notions leading up to integral currents on a subset U of $\mathfrak{R}^2$. Two currents, for example, will be denoted $I_2(U)$. Let δ be a differential form defined on U with compact support, and T be a distribution on δ (i.e. a bounded linear functional on the forms $\Lambda^2(U)$). The boundary of T, ∂T, can be easily defined on one forms δ by $(\partial T)(\delta) = T(d\delta)$, where d denotes exterior derivative. If T and ∂T are continuous with respect to the L^∞ norm on forms, we obtain the normal currents; these can be made into a Banach space using duality. Indeed, let $M(\delta)$ be any norm on 2 forms and let $M(T)$ be the dual norm

$$M(T) = \sup_{M(\delta) \leqslant 1} T(\delta)$$

Let the normal current norm $N(T)$ be defined as

$$N(T) = M(T) + M(\partial T)$$

A special norm of forms, equivalent to the L^∞, called comass (Bombieri's book mentioned in the Bibliography]) has a dual norm called *mass* in which the two-dimensional area of T (in case $T(\delta) = \int_T \delta$) coincides with the two-dimensional area of a Simplex T. Not all normal currents can be obtained by integration on sets however. Thus it is necessary for our purpose to restrict attention to integral currents (i.e. those currents that can be approximated in mass norm by finite polyhedral chains and their deformations under Lipschitz continuous mappings). By this restriction, we shall be able to establish the compactness, closure, and regularity results needed for the application to Poincaré's problem.

We are now in a position to reformulate Poincaré's isoperimetric variational principle in terms of integral currents. First we note that all the above definitions can be carried over to Riemannian manifolds invariantly. In this way, Poincaré's isoperimetric problem can be written

(I$\tilde{\text{I}}$) Find the current $T \varepsilon \, l_2(M)$ such that

(5) $$\inf_{T \varepsilon \Delta} M(\partial T) = \min$$

where $\Delta = \left\{ T \, | \, T \varepsilon \, l_2(M), \, 2\pi = \int_T Kd \parallel T \parallel \right\}$. Here $l_2(M)$ denotes the set of integral 2-currents on M.

Now we list (without proof) the advantages of utilizing the formulation (II) of Poincaré's problem.

1) The Mass norm in $l_2(M)$ is lower semicontinuous, with respect to "flat" convergence.

2) With respect to this very weak notion of convergence ("flat"-convergence of Whitney) standard closure and compactness theorems allow us to prove the extremal of (I$\tilde{\text{I}}$) above is actually attained.

3) Smoothness and simpleness of the extremal can be established in one blow by applying standard regularity techniques of integral currents.

4) A new notion of stability based on perturbation of metrics can be established, for simple closed geodesics on M of minimal nonzero length.

5) Geometrical intuition concerning curves on manifolds can be utilized coherently with the associated analysis.

First we separate the problem into two distinct parts: existence and smoothness except that for integral currents *smoothness also implies simpleness of the extremal. Here "simple" means the extremal has no self intersections.* The existence part is solved by compactness closure and semicontinuity arguments plus in this case: topological arguments based on homology groups. Indeed since cohomology enters analysis naturally via differential forms, homology groups arise naturally by duality for integral currents.

In the conclusion of this analysis, even the simplest geometric one-dimensional problem, when looked at globally, requires new ideas. Explicit solutions are, of course, out of the question. Some quantitative analysis must be carried out which is logically consistent and at the same time leads to an overview of the problem. This is the virtue of the approach discussed above. Here are a few proofs:

<u>Proof that</u> $\inf\limits_{T \varepsilon \Delta} M(\partial T) > 0$, where

$$\Delta = \left\{ T \mid T \varepsilon l_2(M), \int Kd \parallel T \parallel = 2\pi \right\}$$

Notice that no explicit lower bound is given. This result does not follow from classical isoperimetric inequalities. For example in classic research of Huber , we find on an obvious notation,

$$L^2(\partial T) \geq 2\pi A(T) \{ 1 - \frac{1}{2\pi} \int_T Kd \parallel T \parallel \}$$

So that $\inf M(\partial T) \geq 0$.

We prove our inequality using an argument by contradiction, assuming the following compactness and closure theorem for integral currents: a) if a flat limit of integral currents is normal, then the limit is also integral; b) bounded sets of integral currents are precompact in the flat topology.

Now both theorems utilize the notion of flat convergence. There are various equivalent definitions of the notion of flat norm for integral currents; for our purpose the flat norm of $T \varepsilon I_k(V)$

$$F(T) = \inf_{S \varepsilon I_{k+1}(V)} \{M(T - \partial S) + M(S)\}$$

Notice that if $T \varepsilon I_2(V)$ with V a two-dimensional manifold $I_3(V) = \{0\}$ so the flat form coincide with mass norm. We now utilize these facts. Let $T_n \varepsilon \Delta$ be a minimizing sequence for (II) and assume $0 = \inf_\Delta M(\partial T)$. Then $N(T_n) = M(T_n) + M(\partial T_n)$ is uniformly bounded. We have $M(\partial T_n) \to 0$ while

$$M(T_n) = \int d \| T_n \| \leq \int \left[\frac{K}{\inf K} \right] d \| T_n \|$$

<u>Proof</u> <u>that</u> $\inf_{T \varepsilon \Delta} M(\partial T)$ <u>is attained.</u> Again this proof uses the lower semicontinuity mass and the compactness and closer theorem for integral currents. Let $T_n \varepsilon \Delta$ with $M(\partial T_n) \to \inf_\Delta M(\partial T) > a$ (say). Again $N(T_n) = M(\partial T_n) + M(T_n)$ is uniformly bounded. By utilizing the argument in the above paragraph we deduce $T_n \to T$ first in flat norm and then in mss norm (since T_n is maximal dimensional). Moreover, by lower-semicontinuity

$$M(\partial T) = a$$

and by utilizing closure results we get $T \varepsilon \Delta$. Here we have also used the fact that Δ is closed under mass convergence.

Theorem (Berger and Bombieri)   The Poincaré isoperimetric problem for C^3 smooth ovaloids always has a smooth solution C without self-intersections. If the metric g does not differ too much from constant curvature, (the explicit analytical restriction is deleted here) then

(1) C is connected and is the desired closed simple geodesic of shortest length.

(2)   C is stable in the sense of Hausdorff distance with respect to C^3 perturbations of the metric g.

There are basically four steps in the proof of this theorem that have possible generalizations to more general contexts.

(1) To prove the infimum C is attained and is different from 0, i.e. C ≠ point, a point geodesic, the trivial solution in this case.

(2)   To show the minimizing curve   C   has no self-intersections.

(3) To show C is smooth.

(4) To study the stability of C under a change of metric.

Comments on the Theorem

Step (1) is carried out using the theory of two-dimensional integral currents on the two-dimensional manifold V, denoted by $I_2(V)$,using the style of argument of the direct method of the calculus of variations sketched above .

Step (2) utilizing the theory of integral currents follows directly from regularity results.   This is a great advantage of geometric measure theory for this problem.

Step (3) requires recent arguments of Bombieri (see his recent book mentioned in the Bibliography).

Step (4) uses the sharp isoperimetric inequality for curves on the standard sphere.

Section 3.2   Gauss Curvature and Its Extensions

Problem 2   Two-Dimensional Problems

We now turn to analyze a question concerning Riemannian metrics on compact manifolds. This question is also global in nature. The question we have in mind is as follows:

Question: How does one deform a given metric structure g to the simplest possible metric on the given compact manifold M?

This question was posed in the nineteenth century by researchers who studied Riemann surfaces. In fact, for compact two dimensional manifolds, an analogue of the Riemann mapping theorem was found, namely :

Complex function theory and the topological constructions associated "universal covering spaces" showed that such manifolds denoted (M,g) are conformally equivalent to (M, g_c), where (M, g_c) denotes the manifold M with a Riemannnian metric of constant Gaussian curvature c defined on M.

We now raise the equivalent question:

(π_1): Is it possible, analytically, by a conformal deformation to deform a given Riemannian metric structure g to one of constant Gauss curvature K on M?

This question may seem a little strange since we have specified the use of deformation and th use of the Gaussian curvature K for a compact two-dimensional manifold. However, it is exactly this curvature that has the invariant properties we seek.

The key idea is that the geometric question involved does not depend on local coordinate charts because such a question involving Gaussian curvature is intrinsic to the geometry involved. In addition to all this the Gaussian curvature satisfies an important invariant relationship independent of the particular metric g defined on M (i.e., a "conservation law") that has already been described in Problem 1 above.

For any given compact, smooth two-dimensional Riemannian manifold (M, g) with associated topological properties (as in the case of the "geodesic" problem of Poincaré) we have in mind the Gauss-Bonnet result that can be phrased as follows

(*)
$$\int_M K(x)dx = 2\pi \chi (M)$$

This remarkable equation relates the geometric properties of M, namely, the integral of the Gauss curvature, called in Latin "integra curvatura," to its topological properties given by the Euler-Poincaré characteristic $\chi(M)$.

This equation also shows what sort of constant curvature functions K are possible for given compact manifold M. To begin we might ask the naive question: Is it possible to remove the curvature of the deformed metric completely? That is, can we make the constant K identically zero? The partial   answer follows immediately from (*). Indeed, this integra curvatura formula (*) holds for any curvature function K(x) on M. In particular, if K(x) is the constant zero, then the left hand side of (*) must be zero, and thus $\chi(M) = 0$.   This provides a necessary condition for the conformal   deformation to remove curvature from the original metric g completely. As we shall soon see, we shall need a little analysis to prove that this topological condition is also sufficient.

In the same way a manifold admits a metric of constant positive curvature only if $\chi(M) > 0$. The same formula yields the possibility of constant negative curvature metrics that can only be found if $\chi(M)$ is negative.

The other basic word used in our question is the idea of conformal deformation. Conformal deformation is crucial in relating the complex geometry of Riemann surfaces to the real Riemannian geometry of two-dimensional manifolds. Since we wish to find results that have a meaning in the case of Riemann surfaces, we find it important to restrict attention to conformal deformation. Analytically, this simply means that the deformed Riemannian metric g' is related to the original Riemannian metric g by a positive scale factor exp h, and we write

$$g' = (\exp h)g$$

The partial differential approach   For two-dimensional Riemannian manifolds M the usual approach to (π_1) via complex function theory requires the introduction of the universal covering surface $\tilde{M}$ of M. Thus it is useful to give (even in complex dimension one) an alternate approach to the construction of metrics of constant Gauss curvature on nonsimply connected manifolds that (a) does not depend on the covering surface $\tilde{M}$ and (b) can be extended to the higher dimensional case where the complex function theory approach breaks down.

This approach that is advocated here, makes use of a nonlinear Dirichlet principle for the solution of a semilinear elliptic partial differential equation defined on the compact manifold (M, g). Interestingly enough, this equation's solvability could never be achieved by Liouville's method of "quadrature," but requires a more sophisticated existential argument.

To derive the associated partial differential equation we begin by saying that two C^∞ metrics g and $\tilde{g}$ are (pointwise) conformally equivalent if there is a C^∞ function σ defined on M such that $\tilde{g} = e^{2\sigma}g$. Then in terms of isothermal parameters (u, v) locally defined on M, with $g = \lambda(u, v)[du^2 + dv^2]$ the Gauss curvature function K of M can be written

$$K = -(\frac{1}{2\lambda})\{(\log \lambda)_{uu} + (\log \lambda)_{vv}\}$$

Consequently an argument via intrinsic differential geometry shows the deformed Gauss curvature $\tilde{K}$ of $\tilde{g}$ can be written in terms of the Laplace–Beltrami operator Δ on (M, g) as

$$\tilde{K} = e^{-2\sigma}\{K - \Delta\sigma\}$$

Notice this equation has an invariant meaning on (M, g), and so, it can be written independent of local isothermal parameters.

Thus to find a (pointwise) conformally equivalent metric $\tilde{g}$ with Gauss curvature $\tilde{K} = $ const, one must find a smooth function $\sigma \in C^\infty(M)$ that satisfies the partial differential equation

$$\Delta\sigma - K(x) + \tilde{K}e^{2\sigma} = 0 \; ; \quad \tilde{K} = \text{const.}$$

This equation can be written in the form (A) mentioned at the very beginning of this chapter. Indeed, we write this equation as

(*) $$\Delta\sigma + \tilde{K}e^{2\sigma} = K(x)$$

Thus, in terms of the form (A) mentioned above, grad $F(\sigma)$ coincides with the left-hand side of this equation, whereas the inhomogenous term $g = K(x)$, i.e. in this case the given Gauss curvature of the undeformed metric.

Now we invoke the ideas of Chapter 2, namely, the nonlinear Fredholm alternative for equations of the form (A). As mentioned earlier the simplest case occurs precisely when we eliminate all curvature $\tilde{K}$ from the deformed metric. We know a necessary condition for the deformed metric to have Gauss curvature identically 0, is that the Euler characteristic of M vanishes. We now prove

Theorem A sufficient condition for the conformally deformed metric to have Gauss curvature identically 0 is that $\chi(M) = 0$, i.e. that the genus of M is 1.

Proof To prove the sufficiency of the stated condition, we set the deformed Gauss curvature identically equal to 0 in equation (*). Thus, the nonlinear partial differential equation (*) reduces to the simple linear equation

$$\Delta\sigma = K(x)$$

Thus the problem becomes a well-known linear Poisson equation. In fact, the linear Fredholm alternative applies yielding a necessary and sufficient condition for the smooth solution of this equation over (M, g). This condition is easy to state because the Laplace-Beltrami operator Δ on the compact manifold (M, g) is self-adjoint in the appropriate Hilbert space setting and has a one-dimensional kernel, namely, the constant functions. (The reader can easily supply the missing details here). Thus the condition we need is that the constant function 1 is orthogonal to the function $K(x)$ in the L_2 sense over (M, g), i.e.

$$\int_M K(x)dV = 0$$

Now the Gauss-Bonnet theorem applies yielding that this last equation is precisely the condition that the Euler-Poincaré characteristic of M vanishes, i.e. $\chi(M) = 0$ as required.

Now we proceed to the more general nonlinear case of the deformed curvature not vanishing identically. In this case a first observation is that the Gauss-Bonnet formula implies that

(I.6') $\text{sgn } \tilde{K} = \text{sgn } \chi(M)$

Moreover, we require a nonlinear generalization of the classical Fredholm alternative since the nonvanishing of the deformed curvature makes the problem (*) truly nonlinear. In this section we study the case obtained by assuming that $\chi(M)$ is negative, and as an alternative to arguments using covering spaces we prove

Theorem   Under the assumption   that the Euler-Poincare characteristic of M is negative as   above, (M, g) can be conformally deformed to a Riemannian metric of Gaussian curvature -1. In fact the assumption is the desired necessary and sufficient condition for such a conformal deformation.

Proof   Our argument is based on two fundamental steps: first, that a generalized solution of equation (*) exists in the Sobolev space $W_{1,2}$ (M,g). We obtain this solution by simply minimizing an associated functional over this Sobolev space. The second part of the argument is a regularity result stating that the minimum obtained in the first part is smooth enough to yield the given conformal deformation.   This regularity part of our argument follows from the standard regularity theory of linear elliptic equations and is sketched in the problems section. Thus in this book, it suffices to carry out the minimization argument based on our observation that the equation (*) has the form (A) mentioned in the beginning of this chapter.

In this case the associate functional has the form

$$F(u) = \int_M \frac{1}{2}|\nabla u|^2 + \frac{1}{2}e^{2u} + K(x)u$$

Notice that this functional is bounded below on the Sobolev space $H = W_{1,2}(M, g)$. This follows simply by noting that the second term in the integrand dominates the third term, since the manifold M is compact. Then, it is necessary to show that the inf of $F(u)$ over H is attained.   This result follows from general principles.   It is merely necessary to show that the functional is coercive, i.e. that $F(u) \to \infty$ as $\|u\|_H \to \infty$ and has the lower semi- continuity property with respect to weak convergence on H.   Assuming the coerciveness property for the moment, let us examine the argument from first principles. First we know the real <u>valued</u> <u>functional</u> $F(u)$ <u>bounded</u> <u>from</u> <u>below</u> as u varies over H. This is clear because the only negative term in $F(u)$ is linear in   u   and so can be overwhelmed by the positive exponential term in the above equation.

Thus $\inf_H F(u) = m$ is finite.

Now let $\{u_n\}$ be any minimizing sequence in H with the property that $F(u_n) \to m$. By the coerciveness property of $F(u)$ we conclude that the sequence of norms in H, $\{\|u_n\|\}$ is uniformly bounded. Thus by a basic property of Hilbert spaces, $\{u_n\}$ has a weakly convergent subsequence which we relabel $\{u_n\}$. Moreover, we may assume $\{u_n\}$ has the weak limit u_∞. (As an aside we note it turns out that u_∞ is the generalized solution to the geometric problem). Now we show that $F(u_\infty) = m$ using an argument by contradiction. Indeed, suppose otherwise, then $F(u_\infty) > m$. Note $F(u_\infty)$ cannot be less than m, since m is the infimum of F. But, by the lower semicontinuity property of Lebesgue integration, (Fatou's theorem, for example)

$$\underline{\lim} \; F(u_n) \geq F(u_\infty)$$

But by construction $\lim F(u_n) = m$. So we conclude $m \geq F(u_\infty)$. This last argument shows that the infimum of $F(u)$ over H is actually achieved at $u = u_\infty$, so u_∞ is the desired absolute minimum.

As the final step we demonstrate the coerciveness of $F(u)$ over the Hilbert space H, using equation ($\bullet$) below.

To this end, we decompose the function u (uniquely) into its mean value u_m over (M, g) (u_m = a constant), and its part u_0 of mean value 0 over (M, g). Thus

$$(\bullet) \quad F(u_0 + u_m) = \int_M \frac{1}{2} |\nabla u_0|^2 + \frac{1}{2} e^{2u_m} e^{2u_0} - \int_M K(x)u_0 + \int_M K(x)u_m$$

As u_m is a constant, this last integral can be simplified via Gauss Bonnet as

(**)
$$\int_M K(x)u_m = u_m\chi(M)$$

Since $u_0 \perp u_m$ in the Hilbert space H,

$$\|u\|_H^2 = \|u_0\|_H^2 + \|u_m\|_H^2$$

Thus we need only consider the two possibilities (i) $\|u_0\|_H \to \infty$ for arbitrary $\{u_m\}$ and (ii) $\|u_m\|_H \to \infty$ while $\|u_0\|_H$ remains finite. In the first case, case (i), since u_0 satisfies the Poincaré inequality

$$\|\nabla u_0\|_{L_2} \geq c\|u_0\|_{L_2}$$

for some absolute constant c, one simply ignores the second positive term in the integrand, to show using (*) that $F \to \infty$ as $\|u_0\|_H \to \infty$. Here one simply uses the Cauchy-Schwarz inequality. In case (ii), we first let $u_m \to +\infty$ while $\|u_0\|_H$ remains finite. Then we note the exponential term in the integrand for (*) implies $F \to \infty$. On the other hand, if $u_m \to -\infty$, (*) coupled with (**) and the fact that by assumption $\chi(M) < 0$ imply $F \to \infty$ as $u_m \to -\infty$.

Corollary The manifold (M, g) admits a (pointwise) conformally equivalent metric of

(i) constant negative Gauss curvature if and only if the Euler characteristic of M, $\chi(M) < 0$, and
(ii) zero Gauss curvature if and only if $\chi(M) = 0$.

In case $\chi(M) < 0$, the appropriate class of competitors in the minimization problem turns out to be the Sobolev space $W_{1,2}(M, g)$ consisting of all functions $u(x)$ such that $u(x)$ and $|\nabla u|$ are in $L_2(M, g)$. In case $\chi(M) = 0$, the class C may be chosen to be the closed subspace of $W_{1,2}(M, g)$ consisting of those functions of mean value zero over M.

A gap in this approach consists in finding a proof of the case of simply connected compact manifolds by partial differential equations. This difficulty can be traced to the fact that if we attempt to solve (*) with $\tilde{K} > 0$, any critical point of $I(u)$ will necessarily be a saddle point of the functional $I(u)$. To demonstrate this, suppose σ is a relative or absolute minimum of $I(u)$ with $\tilde{K} = 1$. Then we shall find a contradiction to this by showing that for any (nonzero) constant c,

$$I(\sigma + \varepsilon c) < I(\sigma) \quad \text{for } \varepsilon \text{ sufficiently small.}$$

Indeed

$$\delta^2 I(\sigma, c) = \frac{d^2}{d\varepsilon^2} I(\sigma + \varepsilon c)|_{\varepsilon=0} = -2c^2 \int_M \exp 2\sigma < 0$$

so that

$$I(\sigma + \varepsilon c) - I(\sigma) = \frac{1}{2}\varepsilon^2 \delta^2 I(\sigma, c) + 0(\varepsilon^3) < 0$$

Consequently σ cannot be an absolute or relative minimum for $I(u)$. However the simply connected case for compact two-dimensional manifolds is exactly the case treated well by complex analysis. This case is described briefly in the sequel.

Section 3.3   Manifolds of Constant Gauss Positive Curvature

As stated above, the problem just discussed is closely related to the natural problem of finding a metric with the "simplest" curvature properties on a given complex manifold. We now give a small historical survey of ideas on this topic from the nineteenth century and early part of this century before the methods of nonlinear analysis were known.

The topic in complex dimension one can be traced back to Monge and Liouville, who formulated and solved a local version of the problem. The actual global problem reached paramount importance in the late nineteenth century with the work of Poincaré and Klein on the uniformization of algebraic curves. The notion of universal covering surface (introduced by H. A. Schwarz) and the associated Riemann mapping theorem for simply connected Riemann surfaces proved sufficient to solve the uniformization problem. When coupled with the Gauss-Bonnet theorem, these results also yielded complete results on the construction of constant Gauss curvature metrics on any Riemann surface.

Until recently, analogous results on higher dimensional complex manifolds have been noticeably lacking. Perhaps the principal reasons for this situation were (a) there is no known sharp analogue for higher dimensional algebraic varieties as yet of the uniformization theorem for curves; (b) the generalized Gauss-Bonnet theorem for higher dimensional complex manifolds M does not seem useful in restricting the sign of a possible constant curvature metric by the topology of M, alone, as in the Riemann surface case.

The problem in complex dimension one

A. Monge's Problem   Perhaps the earliest study of the construction of metrics of constant curvature is due to Monge, who posed the following "local" problem:

Under what circumstances does the metric $ds^2 = \lambda(x, y)[dx^2 + dy^2]$ have constant total (i.e. Gauss) curvature K?

This problem was solved by Liouville and his solution can be found in a Note in a late edition of Monge's classic <u>Application de analyse à geometrie</u>. Setting $u = \log \lambda$, Monge observed that $u(x, y)$ must satisfy the partial differential equation

(I.1) $$\Delta u = -2Ke^u \quad \text{where} \quad \Delta = \frac{\partial^2}{\partial x^2} + \frac{\partial^2}{\partial y^2}$$

Liouville was able to integrate this equation completely. In fact he found the general solution to Monge's problem could be written in the form

(I.2) $$ds^2 = 4a^2 \exp(2U)\{1 \pm \exp 2U\}^{-2}\{dU^2 + dV^2\}; \quad (K = \pm \frac{1}{a^2})$$

where U and V are the real and imaginary parts of an <u>arbitrary</u> complex-valued function $\zeta(z)$ with $z = x + iy$, so that

(I.3) $$dU^2 + dV^2 = \zeta(z)\bar{\zeta}(z)\{dx^2 + dy^2\}$$

The function $u(x, y)$ defined by (I.2) – (I.3) satisfies (I.1) can be proven. The fact that it represents the general solution of (I.1) is essentially read off from the fact that $u(x, y)$ contains the two arbitrary functions U and V. Unfortunately Liouville's solution is not suitable for a problem more global than Monge's. Indeed, for example, it is easily shown that, for K < 0, no smooth solution of (I.1) can be defined on $\mathfrak{R}^2$.

B. Relation to the uniformization theorem   Nonetheless, a few decades later the problem of determining globally defined metrics of constant Gauss curvature on a Riemann surface arose in the work of Klein and Poincaré on the uniformization of algebraic curves.   This uniformization problem consists of finding a parametrization x = x(t) for the curve defined by polynomial equation F(w, z) = 0, where w and z are complex variables and the parameter t is so chosen as to vary over a simply connected domain of the complex plane. It was found that the uniformization problem could be solved by finding on every compact orientable Riemannian 2-manifold (M,g) a metric ğ of constant Gauss curvature leaving the complex structure of M unchanged.   Indeed, assuming this metric g is known, then the argument proceeds as follows: On an appropriate compact Riemann surface S the relation F(w, z) = 0 can be written w = f(z). Now if we can represent S as the quotient of a domain D in the complex plane by a discontinuous group Γ (acting without fixed points), then the canonical surjection mapping σ : D → D/Γ = S is easily shown to be analytic and single-valued. Thus z = σ(t) and w = f(σ(t)), for t ∈ D, determines the desired uniformization. Now this representation of S, equipped with the metric g by D/Γ is precisely the content of the Clifford-Klein space problem for two-dimensional manifolds of constant Gauss curvature which was solved at the end of the nineteenth century. Thus, the problem is reduced to finding the metric ğ.

Unfortunately, this approach to the uniformization problem was not completed by Poincaré since the necessary analytic tools had not been sufficiently developed in his time. In fact, complete reversal of the two problems occurred; by utilizing H. A. Schwarz's notion of covering space the topological difficulties associated with the nonsimple-connectedness of the Riemann surface associated with $f(x, y) = 0$ were surmounted by Koebe and Poincaré in 1907. The resulting Riemann mapping theorem for simply-connected Riemann surfaces was then shown to yield both the proof of the uniformization of algebraic curves and the existence of the constant curvature metric $\tilde{g}$ mentioned above.

We illustrate this approach via complex function theory in the simplest case in which (M, g) is a simply connected compact orientable Riemannian 2-manifold, and $S(M)$ is the corresponding Riemann surface. One seeks a single-valued (nonconstant) meromorphic function f defined on $S(M)$ that takes the value $z = \infty$ exactly once. Then since a single-valued (nonconstant) meromorphic function takes on all values the same number of times f can be considered as a complex-analytic homeomorphism of $S(M)$ onto the complex sphere. Consequently f determines a conformal diffeomorphism between (M, g) and the sphere (S^2, g_1) with the canonical metric of constant Gauss curvature 1. In order to find the meromorphic function f defined on $S(M)$ it is of great interest to use the Dirichlet principle (in a form due to Hilbert). We choose a point p on M with local parameter z; then we construct a harmonic function U which is smooth everywhere except at some neighborhood $N(P)$ where it behaves like $Re(1/z)$. This harmonic function U is found by minimizing the Dirichlet integral over all C^1 functions in $M - N(p)$ and which behave like U in $N(p)$. (Moreover the differential $dU + i * dV$ is meromorphic on $S(M)$ with the singularity $d(1/z)$ at p). Thus we can state

(I.4) **Theorem**   A compact orientable 2-manifold (M, g) can be given a conformally equivalent metric $\tilde{g}$ of constant Gauss curvature unity if and only if M is simply connected.

The necessity of the simple-connectedness of M   for the validity of this theorem is an immediate consequence of the Gauss-Bonnet formula, since in the case of complex dimension one the Euler characteristic $\chi(M)$ of M is positive if and only if M is simply connected.

Section 3.4   Mean Curvature

Problem 2°   Find Two-Dimensional Surfaces of Constant Mean Curvature

A famous problem in global differential geometry can be phrased:

Find compact surfaces M in $\mathfrak{R}^3$ of constant mean curvature.

Here a compact surface M in $\mathfrak{R}^3$ means that the surface has no boundary. The simplest such example is the standard sphere defined by the single equation $x^2 + y^2 + z^2 = 1$. Heinz Hopf showed that such spheres are the only possibilities for M in the case of genus 0. Hence the question arose,

What happens in the case of higher genus? This question was outstanding for many years until, in 1984, H. Wente found many distinct compact surfaces of genus one using a geometric argument coupled with results from nonlinear partial differential equations He analyzed, for example, the nondesingularization phenomenon as $\lambda \to 0$ for the nonlinear eigenvalue problem

(1) $\Delta u + \lambda \sinh u = 0$ in $D \subset \mathfrak{R}^2$

$$u|_{\partial \Omega} = 0 \quad \text{on } \partial D$$

Here D is a rectangle in $\mathfrak{R}^2$. The discussion here is a modification of an article of my colleague Professor J. Spruck.

To see how the equation (1) arises write the first and second fundamental forms of M in the standard way

$$I = Edu^2 + 2Fdudv + Gdv^2$$
$$II = edu^2 + 2fdudv + gdv^2$$

Taking the lines curvature of M as coordinate axes we may assume $F = f = 0$. In fact we let M have a conformal representation so that

$$ds^2 = E |dz|^2$$

so utilizing our previous formulae for the Gaussian curvature K in terms of the first fundamental form, we find the following equation for K in terms of the function E defined above

(2) $K = -2E^{-1}[\Delta \log E]$

where Δ is the Laplace operator. The Mainardi–Codazzi equations then imply that, if H denotes the mean curvature of M,

(3) $4E^2(H^2 - K) = \lambda$ $(\lambda = \text{constant})$

Substituting (3) into (2) one finds, assuming H is a positive constant,

$$\Delta \log E + 2EH^2 - \lambda^2(2E)^{-1} = 0$$

Define the new variable $u = u(x, y)$ by the formula

$$E = \lambda(2H)^{-1}e^u$$

Thus one has the resulting nonlinear eigenvalue problem

$$\Delta u + 2\lambda H \sinh u = 0$$

To obtain the desired nonlinear problem (1) we simply set $H = 1/2$ in the above equation.

The Dirichlet boundary condition

$$u|_{\partial\Omega} = 0$$

arises by generating the surface M by reflections across the rectangle with sides of length a and b, say. The function u so defined, thus is extended to a doubly periodic function on the plane. In order that such a doubly periodic function gives rise to a doubly periodic conformal representation of the required surface, Wente had to solve the problem of "periods." He found this could be accomplished by choosing the function u, described above, as a non-negative solution of the nonlinear Dirichlet problem mentioned above.

The analysis of (1) to find M required Wente to study the positive smooth solutions of (1) as $\lambda \to 0$. He found the following <u>nonlinear desingularization phenomenon, as was described in Chapter 2</u>.

Theorem   As $\lambda \to 0$ the positive solutions of (1) tend to a constant multiple of the the Green's function for the operator Δu on the rectangle D with zero Dirichlet boundary conditions.

Sketch of Proof   In order to analyze (1), we consider the isoperimetric characterization of the positive solutions of (1)

(π_R) $$\min_{\Sigma_R} \frac{1}{2} \int_D |\nabla u|^2$$

Here, for R fixed, positive and large

$$\Sigma_R = \{u | u \in \dot{W}_{1,2}(D), \ R = \int_D \cosh u\}$$

Using the idea of the solution diamond diagram of the Appendix 3 at the end of Chapter 1, we know that the critical points of (π_R) give rise to positive solutions of (1) for $\lambda = \lambda(R)$. Note here that the positivity property of the solution u desired does require additional argument. The key fact used here is that the isoperimetric problem is left unchanged when the function u is replaced by its absolute value. Consequently, a minimizing sequence for the problem can, without loss of generality, be replaced with nonnegative functions, namely the absolute value of the functions of the minimizing sequence. To analyze the dependence of λ as a function of R we let $R \to \infty$, then we show ina fairly straightforward way that $\lambda(R) \to 0$. As in the example of the nonlinear phenomenon of Section 2.4, the product $\lambda(R)\sinh u_R \to c\delta(x)$ in the space $\dot{W}_{1,p}(R)$ for $p < 2$ (for some constant c). Thus the solution of (π_R) for R very large tends to a multiple of the desired Green's function.

Section 3.5   Simple Riemannian Metrics

Problem 3   Higher-Dimensional Problems

We now consider problems concerning the simplest possible metrics on higher-dimensional Riemannian manifolds (M, g) by considering the natural extensions of the two-dimensional case considered above.

The most direct extension of the work discussed in equation (A) at the beginning of this chapter pertains to obtaining constant curvature metrics via conformal deformation. Here a basic dilemma occurs because there is no clear cut direct generalization of the intrinsic invariant Gaussian curvature for higher dimensional Riemannian manifolds. The most direct real valued function measuring curvature on higher-dimensional Riemannian manifolds is called scalar curvature. The associated problem of conformal deformation relative to scalar curvature is known as the Yamabe Problem, because it was the object of an important and original study published by the researcher H. Yamabe in 1960 shortly before his death. The problem can be stated as follows:

Find a smooth, conformal deformation of the compact smooth higher dimensional Riemannian manifold (M, g) to a Riemannian metric of constant scalar curvature.

It is amazing that this problem is equivalent to solving a nonlinear eigenvalue problem for the conformal deformation defined over (M, g). This nonlinear eigenvalue problem can be described by a semilinear elliptic partial differential equation defined on (M, g) where the associated nonlinear eigenvalue is the desired constant scalar curvature. What makes this equation exceptional is that the nonlinearity in the equation is directly related to loss of compactness in the associated Sobolev inequality, relating the L_p norms of a function, the associated norms of the Sobolev space $W_{1,2}$ (M, g)and the precise nonlinearity of the asso ciated isoperimetric variational problem. Thus an extension of the abstract nonlinear eigenvalue problem result of Chapter 2 is required.

For compact, smooth Riemannian manifolds (M^N, g) of dimension $N \geq 3$ we now seek to extend the discussion of **Problem 2 on nonlinear eigenvalue problems**. As a first step, we try to deform the given metric structure on (M^N, g) to one whose scalar curvature is as simple as possible. (As mentioned above, the scalar curvature is only one of many possible higher-dimensional generalizations of the Gauss curvature function). (The scalar curvature R is simply defined as the trace of the Ricci tensor R_{ij}). Once again, we seek a conformal deformation of metrics. Thus we set

$$g' = e^{2\sigma}g$$

A well-known formula in differential geometry shows the deformed scalar curvature R' satisfies the formula

(†) $R' = e^{-2\sigma} \{ R - 2(N - 1) \Delta\sigma - (N-1)(N-2) |\nabla\sigma|^2 \}$

This formula contains an extra term within the curly brackets when the dimension N exceeds 2, that is the cause of considerable difficulty. By setting $\bar{g}' = u^{\frac{4}{N-2}} g$ (when N exceeds 2)with u a smooth strictly positive function defined on M^N so that $\exp(2\sigma) =$

$u^{\frac{4}{N-2}}$ we find that (†) can be rewritten (provided N exceeds 2 as will be assumed throughout the sequel)

(††)
$$R' = u^{-a_N} \{ Ru - b_N \Delta u \}$$

where the constants a_N and b_N are given by the formulae

$$(N - 2)a_N = N + 2, \quad (N - 2)b_N = 4(N - 1)$$

Once again the simplest case occurs when we attempt to set $R' \equiv 0$, i.e. the deformed curvature vanishes identically. As in **Problem 2**, the problem then becomes linear. In (††) with $R' \equiv 0$ becomes the linear elliptic partial differential equation on M^N

(1)
$$b_N \Delta u - Ru = 0$$

the Δ is the Laplace Beltrami operator on (M^N, g). To solve (1) for the geometric problem means finding a smooth, strictly positive function on M^N. Analytically solving this problem for (1) means proving the smallest eigenvalue of the linear operator $Lu = b_N \Delta u - Ru$ on (M^N, g) is precisely zero. By spectral theory it is well-known analytically that this fact is equivalent to requiring

$$0 = \inf_{\Sigma} \int (b_N |\nabla u|^2 + Ru^2)$$

Indeed, the eigenfunction $u(x)$ associated with (1) can be chosen to be strictly positive and moreover, $u(x)$ is a simple eigenfunction. Why?

Here Σ consists of normalized functions $u(x)$ in $W_{1,2}(M^N, g)$ with L_2 norm equal unity. The difficulty with this condition is that

it cannot be phrased in terms of truly geometric invariants as in
Problem 2 above.

Now we inquire what happens to our problem when we attempt
to find nonzero R', i.e. the deformed scalar curvature constant
R' ≠ 0. Our first observation is easy: the problem becomes
<u>nonlinear</u>. But what kind of nonlinear problem is the equation (††)
The answer is easy if we rewrite (††) as

(2) $b_N \Delta u - Ru = R'u^{aN}$

The geometric problem rephrased in terms of nonlinear partial
differential equations on manifolds has become a nonlinear
eigenvalue   problem (a remarkable   analytic simplification).   We
simply consider   the strictly positive solution of (2), $u(x)$, as a
nonlinear eigenfunction and regard the constant R', the desired
deformed scalar curvature constant, as a nonlinear eigenvalue. Thus,
to repeat, we seek a smooth strictly positive solution $u(x)$, as
required for the geometric problem, as a nonlinear eigenfunction.
Indeed (2) and its solution (u, R') is simply a nonlinear eigenvalue
problem, a direct generalization of the **Problem 2** case mentioned
at the beginning of this Chapter.

To understand the nuances of this nonlinear eigenvalue
problem, we simply extend the characterization of the solution via
the calculus of variations as was discussed abstractly in Chapter 2.
This means we reformulate the solution of (2) as an isoperimetric
variational problem.

Written abstractly, this variational problem can be written

$\inf_{\Sigma_1} (Lu, u)$ where Σ_1 is the set of positive functions in $W_{1,2}(M, g)$

normalized by setting

$$(3) \qquad \int_{M^N} u^{c_N} dV = 1 \qquad \text{with } c_N = \frac{2N}{N-2} = 1 + \frac{N+2}{N-2}$$

Here $(Lu, u) = \int_M (b_N |\nabla|^2 + Ru^2) dV$, with exactly the same inner

product as occurred in the earlier case.

Notice the exponent c_N is always greater than 2 for higher-dimensional manifolds, and has no meaning at all in the two-dimensional case. The other crucial fact about the exponent c_N is that it represents a limiting case of the Sobolev inequality in which compactness of the associated imbedding is lost, as mentioned in the Appendix 3 of Chapter 1. For nonlinear eigenvalue problems this fact turns out to cause great difficulties, since in the theoretical development of nonlinear eigenvalue problems in this book and elsewhere compactness is an essential ingredient in key points of the proofs. If compactness prevails, abstract methods can generally be used to solve nonlinear eigenvalue problems in a direct manner. However, when compactness breaks down, such as in the present case, a new subtle element is introduced into the nonlinear eigenvalue problem. (Notice this fact is also true in linear self-adjoint eigenvalue problems.) In the subsequent paragraphs, we shall show that certain direct arguments involving steepest descent techniques not relying on compactness hypotheses can be utilized with some success.

Now we apply nonlinear analytic techniques to this problem. We rewrite the nonlinear eigenvalue problem as a Hilbert space problem in the space $H = W_{1,2}(M, g)$

$$Lu = \lambda Nu \quad \text{where} \quad \lambda = \frac{(Lu, u)}{(Nu, u)}$$

The simplest approach to solve this nonlinear eigenvalue problem is to minimize the quadratic form (Lu, u) over the set Σ_1. This isoperimetric problem has the virtue that the nonlinear eigenvalue R' (in this case) is not mentioned explicitly but rather is regarded as a Lagrange multiplier to be determined a posteriori.

To achieve this minimization procedure, we extend the method of steepest descent described in Chapter 2 to minimization of the functional A(u) over hypersurfaces M defined over a Hilbert space H defined by a level surface of the functional B(u).

Here in the geometric application, we shall choose $A(u) = (Lu, u)$ and $B(u) = \int_M u^{c_N} dV$.

In symbols, $M = \{ u| u \in H, B(u) = 1 \}$ and assume throughout that $B'(u) \neq 0$ for $u \in M$.

Consider the initial value problem

(4)
$$\begin{cases} u'(t) = -\text{grad}_M A(u) \\ u(0) = u_0 \end{cases}$$

where $u_0 \in M$ and $\text{grad}_M A(u) = A'(u) - \lambda(u)B'(u)$

(4')   with $\lambda(u) = (A'(u), B'(u))/ \|B'(u)\|^2$.

Since $(\text{grad}_M A'(u), B'(u)) \neq 0$, as can be verified by a simple computation, $\text{grad}_M A'(u)$ us a C' tangent vector field on the infinite-dimensional manifold M. Then by a slight extension of the standard existence theory for initial value problems mentioned in Chapter 2, to a Hilbert space context, the initial value problem for (4) has the unique solution $u(t)$ existing for small t, and lies on M. Along $u(t)$, a simple computation shows

$$\frac{d}{dt} A(u(t)) = (A'(u(t)), u'(t) \leq 0$$

Moreover, we have the following result.

Theorem   Assume A is bounded from below on M, and $B'(u) \neq 0$ for $u \in M$.   Then the initial value problem (4) exists for all $t \in (0, \infty)$ and moreover

$$\int_0^\infty \| \text{grad}_M A'(u(t)) \|^2 dt < \infty$$

Proof   Suppose the solution exists only on the maximal interval $[0, T)$ with T finite. Then, in that interval, via Cauchy–Schwarz

$$\| u(t_2) - u(t_1) \| = \| \int_{t_1}^{t_2} u'(t)dt \|$$

$$\leq \int_{t_1}^{t_2} \| \text{grad}_M A'(u(t)) \| dt$$

$$\leq (t_2 - t_1)^{\frac{1}{2}} \int_{t_1}^{t_2} \| \mathrm{grad}_M \, A'(u(t)) \|^2)^{\frac{1}{2}}$$

While by above computations

$$A(t_2) - A(t_1) = \int_{t_1}^{t_2} \frac{d}{dt} A(u(t)) dt$$

$$= -\int_{t_1}^{t_2} \| \mathrm{grad}_M \, A(u(t)) \|^2 dt$$

As the functional $A(u)$ is bounded from below M, the above equations imply as $t \to T$, $u(t)$ forms a Cauchy sequence, so that $\lim u(t)$ as $t \to T$ exists in H a contradiction. Consequently T must be infinite. Then the finiteness of the infinite integral follows from the last equation.

Corollary   In addition to the above hypotheses, (1) on M, suppose the sets

$$M_a = \{ u \mid u \in M \quad A(u) \leq a \}$$

are bounded for each finite number a and (2) $\mathrm{grad}_M A$ satisfies the following continuity property: $u_n \to u$ weakly and $\mathrm{grad}_M A(u_n) \to v$ strongly implies $\mathrm{grad}_M A(u) = v$. Then the solution $u(t)$ of the initial value problem (4) has the weak limit u_∞. This weak limit satisfies the equation

$$A'(u) = \lambda B'(u) \qquad \text{with } \lambda \text{ given by (4')}$$

If u_∞ lies on M, it is the desired critical point.

Proof   The above theorem implies there is an increasing sequence t_k with $\text{grad}_M Au(t_k)$ strongly convergent to zero. By hypothesis (1) of the corollary the set $M_{A(u_o)}$ consisting of points u of M with $A(u) \leq A(u_0)$ is bounded. Hence the solution of (4) $u(t)$ has a weakly convergent subsequence $u(t_k)$ with limit u_∞. Moreover, by the above theorem, the convergence of the stated infinite integral implies $\text{grad}_M A(u(t_k))$ converges strongly to zero in H. Thus hypothesis (2) of the corollary $u(t_k)$ converges weakly to u_∞ which implies $\text{grad}_M A(u_\infty) = 0$. Thus u_∞ satisfies the nonlinear eigenvalue problem $A'(u) = \lambda B'(u)$ with λ given by (4'). If u_∞ lies on M, it is indeed the desired solution.

Now for the geometric problem on scalar curvature.

$$2A(u) = (Lu, u) = \int_M (b_N |\nabla|^2 + Ru^2)dV$$

$$\text{with } Bu = \int_M u^{c_N}dV \qquad \text{with } c_N = \frac{2N}{N-2}$$

We first verify the hypotheses assumed above.

Lemma   For the geometric problem just mentioned the functionals A and B are C^2 on H. Moreover,

   (i)   A(u) is bounded below on M,
   (ii) the sets M_k for each finite k are bounded on H
  (iii) $B'(u) \neq 0$ on M.

Proof   The C^2 nature of the functionals A and B is easy since A is quadratic and B has as integrand a power of the argument u for which Sobolev's inequality applies.

Moreover, to verify (i) and (ii), we note on M

$$2A(u) = \int_M b_N [|\nabla|^2 + u^2] + (R - b_N)u^2$$

$$> \cdot b_N \|u\|^2_{1,2} + const$$

Here Hölder's inequality has been applied to bound the second term by a constant via the definition of M by (3) above. To verify (iii), we merely note for any u, v ∈ H by definition

$$(B'(u), v) = \frac{d}{d\varepsilon} B(u + \varepsilon v)|_{\varepsilon=0} = const \int_M u^{N-1} v$$

To complete the geometric problem following the 1982 paper of A. Inoue (Tohoku Math. Jour., vol. 34, pp. 499-507), we modify the abstract method just described by considering any minimizing sequence u_k for the geometric isoperimetric problem

$$\mu = \inf_{\Sigma_1} A(u)$$

and its improvement by the steepest descent technique just described

$$u'_k(t) = grad_M A(u(t)) \qquad u_k(0) = u_k$$

to get an improved minimizing sequence.

Thus we can find a positive sequence $\varepsilon_k \to 0$ with

$$\| \text{grad}_M A(u_k(t_k)) \| \leq \varepsilon_k$$

Now $\| u_k(t_k) \|$ is uniformly bounded by the property (ii) of $A(u)$ on M, so we may assume, by passing to a subsequence that $v_k = u_k(t_k)$ converges weakly to $u \in H$. Moreover a short computation yields $\lambda(u_k) \to \left(\dfrac{2}{c_N}\right) \mu$. At this point it is necessary to utilize geometric inequalities to get more information on the weak limit u of v_k.

To this end we state

Theorem (On the Best Constant for the Sobolev Imbedding Theorem) (T. Aubin) There is a universal constant $K(N)$ such that for any compact Riemannian M^N and any $\varepsilon > 0$, there is a positive constant $b(\varepsilon)$ for which

$$(\dagger) \qquad \|v\|^2_{c(N)} \leq (K(N) + \varepsilon)\|\nabla v\|^2_{L_2} + b(\varepsilon)\|v\|^2_{L_2}$$

holds for all $v \in H$. Moreover

$$K^{-1}(N) = \frac{N(N-2)\omega_N^{N/2}}{4}$$

where ω_N denotes the surface area of the N-dimensional sphere S^N of radius one.

We shall now show how this inequality $(\dagger)$ can be used to give a partial resolution of the Yamabe problem on constant scalar curvature utilizing our argument in the above paragraphs.   Now let $v = v_k \in M$ in $(\dagger)$ with $k \to \infty$.   Now assume, in order to obtain a contradiction, $v_k \to 0$ weakly in H. Now by taking inner product in the definition of $grad_M A(u)$ with v_k

$$(v_k, grad_M A(v_k)) = -2(Lv_k, v_k) + c_N \lambda(v_k)$$

$$(Lv_k, v_k) = b_N \|\nabla v_N\|^2 + \int R v_N^2$$

tends to $b_N \|\nabla v_N\|^2$.

This result will enable us to show in the abbreviated argument given below that the weak limit, in question, u lies on M and so yields the desired nonzero generalized solution without compactness assumption on B'(u).

First note that it suffices to show that u is not the zero element. Otherwise with $u \neq 0$ by considering the ray cu as c varies over the real numbers, we obtain a contradiction to the extremal properties of u.

Now suppose, as above, $v_k \to 0$ weakly in H and we shall obtain a contradiction for certain Riemannian manifolds M^N by assuming $\mu < N(N - 1)\omega_N^{2/N}$ where ω_N is the surface area of the unit sphere S^N. It can be shown that this result solves the Yamabe problem for nonconformally flat compact manifolds M^N of dimension ≥ 6. For the remaining important class of manifolds there is a variety of arguments that can be given, some that differ substantially from the style of argument given above. In dimensions 3 and 4 , there is an important argument, not discussed here, due to R. Schoen. Here is the result.

Theorem (Aubin)   Suppose $\mu < N(N-1)\omega_N^{2/N}$,   then the weak solution u of the isoperimetric problem (π_1) lies on M and so is the desired weak solution of the geometric problem.

Proof (Inoue)   We use the above construction of the improved minimizing sequence $v_k = u_k(t_k) \in M$ and suppose as above $v_k \to 0$ weakly. By definition

$$\text{grad}_M A(v) = A'(v) - \lambda(v)B'(v)$$

Now $v_k \to 0$ weakly in H and by the Rellich compactness theorem $v_k \to 0$ strongly in L_2. In the above equation the left hand side tends to zero whereas from the right hand side we find

$$\lambda(v_k) \to \frac{2}{c_N}\mu$$

and
$$(A'(v_k), v_k) \to c_N\|\nabla v_k\|^2 .$$

Thus we find
$$a_N\|\nabla v_k\|^2 \to \mu .$$

So we find on substituting in (†) above, for $v_k \in M$

$$1 \le [K(N) + \varepsilon]\frac{\mu}{a_N} + b(\varepsilon)\|v_k\|^2_{L_2}$$

Thus if we suppose $\mu \dfrac{K(N)}{a_N} < 1$ we obtain a contradiction, i.e. $\mu < a_N K^{-1}(N)$.   Since $a_N = 4(N-1)(N-2)^{-1}$, we obtain the desired result.

Section 3.6   Einstein Metrics

A Variational Method for Determining Einstein Metrics on Compact Complex Manifolds

In the Yamabe problem discussed above, we discussed deforming a Riemannian metric on higher-dimensional manifolds without regard to any possible complex structures involved. Indeed, in extending the real two-dimensional analysis given in **Problem 2** to the higher-dimensional case, one notes that the real and complex structures coincide in two real dimensions. Thus, in higher-dimensional cases, one must pay careful attention to distinguish between any actual real and complex structures involved. In this section we deform a metric on a complex compact manifold of higher dimension by means of deformations more compatible with the complex structures involved. Although conformal deformation was appropriate in one complex dimension or equivalently, two real dimensions, such deformation are not compatible with the complex structures on higher-dimensional manifolds we have in mind.

We now attempt to deform a special class of simple Hermitian metrics, known as Kähler metrics g, on a complex compact manifold M to one of constant scalar curvature. In fact, the deformed metric we have in mind will be an Einstein metric first defined by Einstein himself in the early part of this century. Such a metric is defined by stating that the associated metric tensor, here denoted g, is strictly proportional to the Ricci tensor R_{ik}. In the sequel, we shall specialize the complex manifolds studied to be Kähler, and the deformation to be such that the Kähler structure is preserved. There will be many interesting geometric invariants associated with such a deformation, extending the Gauss-Bonnet theorem of one complex dimension as mentioned in the above sections.

Introduction

We now take up a new variational principle for the determination of Einstein-Kähler metrics g' obtained by deforming a given compact Kähler manifold (X, g) of arbitrary complex dimension m, while preserving the Kähler structure throughout the deformation. We show the calculus of variations point of view leads to a unified approach to the problem we study, analogous to the study of constant Gaussian curvature, metrics discussed earlier in this chapter. In particular, starting from a given Kähler metric g on X, we consider Kähler deformations of g of the form

$$g_\phi = g + \partial\bar\partial\phi$$

Here, in terms of local coordinates, $\partial\bar\partial\phi$ denotes the matrix

$$\partial^2\phi/\partial z_i \partial\bar z_k$$

As we shall show, this class of deformation preserves the Kähler structure of X. We seek a Kähler deformation g_ϕ (with real valued globally defined function ϕ) which is Einstein-Kähler, i.e., the associated Ricci tensor R_{ik} is proportional to the associated metric tensor g_{ik}. In terms of tensor equations, we write

$$R_{ik} = \lambda g_{ik} \quad \text{where } \lambda \text{ is a constant.}$$

Such deformations have been studied by numerous authors. E. Calabi first initiated the study of this problem in 1954. He proposed various methods for studying it, one of which involved calculus of variations based on minimizing the L_2 norm of deformed Ricci tensors in the space of Kähler metrics defined by the above deformations. The method we present below, in keeping with the discussions carried out in earlier sections of this book, is substantially different, and consists in writing down the partial differential equations satisfied by the function ϕ associated to the desired Einstein-Kähler metric g_ϕ and then computing, by well-established functional analytic results, the convenient variational principle whose Euler-Lagrange equations are indeed the partial differential equation of elliptic type in question. In addition, we are able to formulate the problem in the Sobolev space $W_{2,m}(X)$ which is adapted to the associated variational principle.   Recently, T. Aubin and S. T. Yau, in prize-winning work, building on previous studies by other distinguished researchers, solved successfully the resulting nonlinear partial differential equation when the associated first Chern class of the manifold X, denoted $C_1(X) \leq 0$. These authors utilized the method of continuity, using elaborate a priori estimates. In this section we try to extend the unified pattern obtained in the real two-dimensional case in **Problem 2** to the higher-dimensional case and to make first steps toward solving the variational problem in significant cases.

The use of calculus of variations is extremely natural from both geometric and physical points of view. Einstein and Hilbert attempted novel variational principles to describe Einstein's field equations in general relativity. However, the elaboration of these techniques was impeded by the lack of progress in the mathematical understanding of the calculus of variations. In differential geometry, the use of variational methods in multi-dimensional problems centers around the work of Yamabe discussed earlier. In his case, the nonlinear partial differential equation for the desired deformation is naturally posed in $W_{1,2}(X)$ and the nonlinearity which occurs is a limit case and corresponds to the continuous but noncompact Sobolev inclusion $W_{1,2}(X) \subset L_p(X)$. In work discussed above, the case of one complex dimension and Gaussian curvature was considered. Amazingly, the same limiting behavior which appears in this work. Moreover, our discussion (cf. **Problem 2**) will become apparent in the problem of Einstein–Kähler metrics, namely exponential nonlinearities, and continuous but noncompact Sobolev embeddings, except we consider the geometric problem in terms of the Sobolev space $W_{2,m}(X)$, where the compact Kähler manifold X has real dimension 2m.

(iv) From now on, we shall suppose X is a compact Kähler manifold (i.e., admits Kähler metrics). Let us give some definitions adapted from the text of Aubin, <u>Nonlinear Analysis on Manifolds.</u> A complex manifold X is a paracompact Hausdorff space such that where two neighborhoods overlap the local coordinates transform by a complex analytic transformation. Such manifolds always admit a Hermitian metric g, an analogue of a Riemannian metric which is compatible with the the complex structure introduced on X. We denote the Hermitian metric on X by

$$g = \sum g_{\alpha\bar{\beta}} \, dz^{\alpha} dz^{\bar{\beta}}$$

Associated to this metric, we have the (1, 1) Kähler form

$$\omega = \sum g_{\alpha\bar{\beta}} \, dz^{\alpha} \wedge dz^{\bar{\beta}}$$

and the volume element

$$dV = \text{const. det} \, (g_{\alpha\bar{\beta}}) dx$$

The first fundamental form of a Hermitian manifold (X, g) is the real $(1, 1)$ form

$$\omega = \left(\frac{i}{2\pi}\right) g_{\lambda\bar{\mu}} \, dz^\lambda \wedge dz^{\bar{\mu}}$$

g is said to be Kähler if the $(1, 1)$ form ω, introduced above, is closed: $d\omega = 0$, where d denotes the exterior derivative. Here we refer to the text <u>Complex</u> <u>Manifolds</u> by Kodaira and Morrow.

The following lemma describes the elements of a given cohomology class of $(1, 1)$ forms. (See the text mentioned above for a proof).

Lemma 1   Let $\chi = ia_{\lambda\bar{\mu}} \, dz^\lambda \wedge dz^{\bar{\mu}}$ be a closed real $(1, 1)$ form on a compact Kähler manifold X. Then χ is cohomologous to zero if and only if there exists $f \in C^\infty(X)$ such that

$$\chi = id'd''f, \quad \text{i.e.,} \quad a_{\lambda\bar{u}} = \partial_{\lambda\bar{u}} f$$

A cohomology class $[C]$ of real $(1, 1)$ forms is said to be positive definite (resp. negative definite or null) if $[C]$ contains a representative $\chi = ia_{\lambda\bar{\mu}} \, dz^\lambda \wedge dz^{\bar{\mu}}$ such that the Hermitian matrix $(a_{\lambda\bar{\mu}})_{\lambda,\mu}$ is everywhere positive definite (resp. negative definite or null). Using Lemma 1, one shows that these three properties are mutually exclusive. Thus it is meaningful to say that the important $(1, 1)$ form associated with the first Chern class $C_1(X)$ has prescribed sign $(> 0, < 0,$ or null$)$.

Kähler manifolds can be characterized locally by saying that the $(1, 1)$ form ω is closed if and only if there is a smooth function f defined locally on X such that

$$g_{\alpha\bar{\beta}} = \partial^2 f/\partial z^\alpha \partial \bar{z}^\beta$$

Let g and g' be two Kähler metrics. The quantities corresponding to g' are indicated with a prime ('). g' is called a Kähler deformation of g if the first fundamental forms ω and ω' are cohomologous; equivalently, according to Lemma 1, if there exists $\phi \in C^\infty(X)$ such that

(1) $$g'_{\lambda\bar{\mu}} = g_{\lambda\bar{\mu}} + \partial_{\lambda\bar{\mu}} \phi$$

or in terms of Kähler forms

(1') $$\omega' = \omega + \frac{i}{2\pi} d'd''\phi = \omega - \frac{i}{4\pi} dd^c\phi$$

We shall write the above equations as

$$g' = g_\phi = g + \partial\bar{\partial}\phi$$

The important point to notice is that a Kähler deformation preserves the Kähler property of the associated Hermitian metric, i.e., if $d\omega = 0$, then $d\omega' = 0$. It will be important in the sequel to find the appropriate invariants that remain invariant under such a deformation. Moreover, notice that, in higher dimensions, Kähler deformations are completely distinct from the conformal deformations mentioned earlier in this chapter.

For g' to define a metric on the manifold X, ϕ must belong to the convex set [U] of admissible functions defined by

(2) $[U] = \{ \phi \in C^2(X);$ the matrix $(g_{\lambda\bar{\mu}} + \partial_{\lambda\bar{\mu}} \phi)$
          is everywhere positive definite}.

If $\phi \in [U]$, we can choose local coordinates adapted to a given point $P \in X$, i.e., such that, in P, the matrix $(g_{\lambda\bar{\mu}})_{\lambda,\mu}$ and $(\partial_{\lambda\bar{\mu}} \phi)_{\lambda,\mu}$ are respectively the identity and diagonal. Then, for every direction λ, $(1 + \partial_{\lambda\bar{\lambda}} \phi)(P) > 0$ and the trace $g^{\lambda\bar{\mu}} g'_{\lambda\bar{\mu}} = m + \Delta\phi$ is positive.

Under condition (1), the volume elements relative to g and g' are related by

$$dV' = M(\phi)dV,$$

where M is the complex Monge-Ampère operator

(3) $M(\phi) = \dfrac{|g'|}{|g|} = \det(\delta^{\lambda}_{\mu} + \nabla^{\lambda}_{\mu} \phi),$

since, in the Kähler metric g, it is a relatively easy computation to verify

$$\nabla^{\lambda}_{\mu} \phi = g^{\lambda\bar{\nu}} \partial_{\mu\bar{\nu}} \phi .$$

It is crucial for the Kähler metric that the components of the Ricci tensor satisfy the following relation

(†) $R_{\lambda\bar{\mu}} = -\partial_{\lambda\bar{\mu}} \log|g|$

where $|g|$ is the determinant of the metric tensor. And we write the Ricci form of the metric g as

(††) $\psi = \left(\dfrac{i}{2\pi}\right) R_{\lambda\bar{\mu}} \, dz^{\lambda} \wedge dz^{\bar{\mu}}$

The element ψ is called the Ricci form associated with g. According to (†) the form ψ is closed and hence defines a cohomology class called the first Chern class $C_1(M)$. Higher Chern classes $C_2(M)$, $C_3(M)$, . . .$C_m(M)$ can be defined for a complex manifold of complex dimension 2m. These higher Chern classes $C_i(M)$, are found by expanding the determinant of the matrix

$$\{ I + \Omega/2\pi i \}$$

where Ω is called the curvature matrix, an $m \times m$ matrix of 2-forms and the Chern class $C_p(M)$ is the term of degree 2p ,apart from an absolute constant. The cohomology class of $C_p(M)$ is independent of the Kähler metric. To find some geometric invariants associated with the Chern classes and the Kähler form ω, we consider:

the cohomology class of the form $C_i(M) \wedge \omega^{m-i}$

Fact   The cohomology classes of the above m+1 forms Θ_i (i = 0, 1, 2 ... m) are independent of the Kähler deformation.

Proof   Consider a Kähler deformation of the Kähler metric g. Then since both the Chern classes $C_i(M)$ and the Kähler forms are closed we find

$$\int_M c_i \wedge \omega^{m-i} = \int_M \{c_i \wedge (\omega + d\alpha)^{m-i}\}$$

Consider the first few cases of this result :

First, the case $\Theta_0 = \int_M C_0 \wedge \omega^m$.

$C_0 = 1$ and ω^m is proportional to the volume form dV_g .
Consequently, $\Theta_0 = $ const. implies the volume of a Kähler manifold
remains constant under a Kähler deformation.

Next, the case $\Theta_1 = \int_M C_1 \wedge \omega^{m-1}$.

In this case, it can be shown that $C_1 \wedge \omega^{m-1}$ is proportional to
the trace of the Ricci tensor of   M, i.e., proportional to the scalar
curvature $R(g)$. Thus $\Theta_1 = $ const. implies that during a Kähler
deformation, the "integra curvatura" remains constant, i.e.

$$\int_M R(g) \, dV_g = \text{constant}$$

Notice that this is a generalization of the Gauss-Bonnet theorem
used earlier in this chapter.

Continuing in this way, one determines $m+1$ invariants
of the Kähler deformation. In fact, Θ_m is proportional to the Euler-
Poincaré characteristic, so that we obtain another extension of the
Gauss-Bonnet theorem for complex manifolds of higher complex
dimension .

Let us also compare Ricci forms ψ and ψ' obtained by Kähler deformation, using $(^\dagger)$ and $(^{\dagger\dagger})$ above.

$$(4) \qquad R'_{\lambda\bar{\mu}} - R_{\lambda\bar{\mu}} = -\partial_{\lambda\bar{\mu}} \text{Log} \frac{|g'|}{|g|} = -\partial_{\lambda\bar{\mu}} \text{Log } M(\phi),$$

so

$$\psi' - \psi = (\frac{i}{4\pi})dd^c \text{ Log } M(\phi).$$

where M is the complex Monge-Ampère operator

$$M(\phi) = \frac{|g'|}{|g|} = \det(\delta^\lambda_\mu + \nabla^\lambda_\mu \phi)$$

In the next few paragraphs we shall find that this Monge-Ampère operator will be of crucial importance for the deformation to an Einstein metric.

2. The Equations Leading to Einstein-Kähler Metrics

A Kähler metric on a Kähler manifold X is said to be Einstein if there exists a real number k such that

$$(1) \qquad \psi' = k\omega', \quad \text{i.e.,} \quad R'_{\lambda\bar{\mu}} = kg'_{\lambda\bar{\mu}}$$

A priori (1) is a system for the unknown metric g'. When X is compact, let us prove that it can be reduced to a single nonlinear partial differential equation of second order, by utilizing the notion of Kähler deformation.

A necessary condition for solving (1) is for $C_1(X)$ to be of prescribed sign. First let us suppose $C_1(X) \neq 0$. The coefficient k will have the sign of $C_1(X)$ and, since a homothety on the metric preserves Ricci tensor, we can search g' such that

(2) $$\psi' = \varepsilon\omega' \quad \text{or} \quad R'_{\lambda\bar{\mu}} = \varepsilon g'_{\lambda\bar{\mu}} ,$$

with ε equal to 1 or -1 according whether $C_1(X)$ is >0 or <0.

Choose a real (1-1) form $\chi = \left(\dfrac{i}{2\pi}\right) \varepsilon g_{\lambda\bar{\mu}} dz^\lambda \wedge dz^{\bar{\mu}}$ belonging to $C_1(X)$ such that the matrix $(g_{\lambda\bar{\mu}})_{\lambda,\mu}$ is everywhere positive definite and thus defines a Kähler metric g on X (since χ is closed) with first fundamental form $\omega = \varepsilon\chi$. Hence $\varepsilon\omega$, as well as ψ, belongs to $C_1(X)$ and, by Lemma 1, there exists $f \in C^\infty(X)$ such that

(3) $$R_{\lambda\bar{\mu}} = \varepsilon g_{\lambda\bar{\mu}} + \partial_{\lambda\bar{\mu}} f .$$

On the other hand, (2) shows that $\varepsilon\omega' \in C_1(X)$; thus ω and ω' are cohomologous and g' must be a Kähler deformation of g,

(4) $$g'_{\lambda\bar{\mu}} = g_{\lambda\bar{\mu}} + \partial_{\lambda\bar{\mu}} \phi ,$$

where the unknown function ϕ satisfies an equation to be exhibited. The above relations imply

$$R'_{\lambda\bar{\mu}} - R_{\lambda\bar{\mu}} = \varepsilon(g'_{\lambda\bar{\mu}} - g_{\lambda\bar{\mu}}) - \partial_{\lambda\bar{\mu}} f = \partial_{\lambda\bar{\mu}} (\varepsilon\phi - f) ;$$

since

$$R'_{\lambda\bar{\mu}} - R_{\lambda\bar{\mu}} = -\partial_{\lambda\bar{\mu}} \operatorname{Log} M(\phi),$$

we have

$$\partial_{\lambda\bar{\mu}} [\operatorname{Log} M(\phi) + \varepsilon\phi - f] = 0.$$

Consequently, the function $\operatorname{Log} M(\phi) + \varepsilon\phi - f$, whose Laplacian is zero, must be constant and, because f is defined up to additive constants, ϕ verifies

(5) $$M(\phi) = \exp(-\varepsilon\phi + f).$$

Now, if $C_1(X) = 0$, start with an arbitrary initial Kähler metric g. Then ϕ, cohomologous to zero, can be written $\left(\dfrac{i}{2\pi}\right)d'd''f$, that is, $R_{\lambda\bar{\mu}} = \partial_{\lambda\bar{\mu}} f$, a substitute to (3). If we search a Kähler deformation $g' = g + \partial\bar{\partial}\phi$ of g with null Ricci tensor, the preceding argument leads to the equation

(6) $$M(\phi) = e^f.$$

Note that with $\varepsilon = 0$, this equation (6) is nonlinear because the function $M(\phi)$ is nonlinear. Nonetheless, equation (6) requires exactly an analogue of the linear Fredholm alternative for its solution, as in the discussions given above.

This is a special case of Calabi conjecture which asserts that, on any compact Kähler manifold (X, g), every element

$$\chi = \left(\frac{i}{2\pi}\right) a_{\lambda\bar{\mu}} \, dz^\lambda \wedge dz^{\bar{\mu}} \in C_1(X)$$

is the Ricci form of some Kähler metric g'. In fact, since χ and ψ
belong to $C_1(X)$,

$$R_{\lambda\bar\mu} - R'_{\lambda\bar\mu} = R_{\lambda\bar\mu} - a_{\lambda\bar\mu} = \partial_{\lambda\bar\mu} f$$

for some function $f \in C^\infty(X)$, and if we search g' as a Kähler
deformation of g, equality (4) yields equation (6).

3. Variational Principles for the Deformation Equations

In complex dimension $m = 1$, $1 - M(\phi) = \Delta\phi$, and the smooth
solutions of $\Delta\phi + f = 0$ are smooth critical points of the functional
$\int_X (\frac{1}{2}|\nabla\phi|^2 + f\phi)dV$. It is natural to ask if this extends in higher
dimensions and if the deformation equations can be written as the
Euler-Lagrange equations of some functionals. For gradient
operators and their associated functionals, the theory of Chapter 2
relative to the Sobolev space $W_{2,m}$ leads one to introduce the
following functional, first for smooth ϕ

$$(7) \qquad J(\phi) = \int_X \int_0^1 \phi\, [1 - M(s\phi)]\, ds\, dV \qquad \text{for} \quad \phi \in C^2(X)$$

then for $\phi \in W_{2,m}(X)$ once $J(\phi)$ is appropriately redefined.

It will turn out that all the operators in (5), (6) and (7) are
gradient operators. In Lemma 2 below, we prove the Gateaux
derivative of $J(\phi)$ is $1 - M(\phi)$. We also define (for comparison
purposes) a generalization of the quadratic form associated with
$1 - M(\phi)$, as follows:

(8)
$$j(\phi) = \int_X \phi[1 - M(\phi)]dV.$$

The development of $M(\phi)$ and the definitions (8), (7) imply that

(9)
$$j = \sum_{k=1}^{m} I_k \quad \text{and} \quad J = \sum_{k=1}^{m} \frac{I_k}{k+1},$$

when I_k is homogeneous in φ of degree $k + 1$ and is the polarization of the following $(k + 1)$-linear form H on $C^2(X)$

(11)
$$I_k(\phi) = H_k(\phi, ..., \phi).$$

Let us rewrite the deformation equations under the form

(12)
$$\Gamma_0(\phi) = [1 - M(\phi)] + e^f - 1 = 0$$

(13)
$$\Gamma_\pm = [1 - M(\phi)] - 1 + \exp(-\varepsilon\phi + f) = 0, \qquad \varepsilon = \pm 1$$

and let us consider the associated functionals

(14)
$$I_\pm(\phi) = J(\phi) - \int_X \phi \, dV - \varepsilon \int_X e^{-\varepsilon\phi + f} \, dV$$

(14')
$$I_0(\phi) = J(\phi) + \int_X (e^f - 1) \phi \, dV.$$

These functionals, directly above, yield variational principles for the desired Kähler-Einstein metric obtained by Kähler deformation from a given Kähler metric g. First we prove, directly below, that smooth critical points of these functionals yield solutions of the associated deformation equations. Next, we give some ideas about the use of the direct method of the calculus of variations on the Sobolev space $W_{2,m}(M, g)$ to obtain critical points of the associated functionals.

Lemma 2   The C^2 critical points of the functional I_+, I_-, or I_0 are the C^2 solutions of the corresponding deformation equation.

Proof   We must only verify that if u and v belong to $C^2(X)$ and $j(t) = J(u + tv)$, then

$$j'(0) = \int_X [1 - M(u)]\, v\, dV.$$

By symmetry of the $(k + 1)$-linear form H_k,

$$\frac{d}{dt} I_k(u + tv)\big|_{t=0} = \text{coefficient of } t \text{ in } H_k(u + tv, ..., u + tv)$$

$$= (k + 1)\, H_k(u, ..., u, v),$$

hence

$$j'(0) = \sum_{k=1}^{m} H_k(u, ..., u, v) \qquad\qquad \text{(by (9) and (11))}$$

$$= \int_X [1 - M(u))\, v\, dV$$

A weak solution of a deformation equation (13) is a critical point of the corresponding functional $I_+, I_-,$ or I_0 in the Sobolev space $W_{2,m}(X)$ (see Proposition 1 below).

We now turn to a discussion of such critical points.

II. Sobolev Space Setting for the Variational Approach to Deformation Equations

A key idea of this book is the actual utilization of appropriate variational principles to discover their associated extremals and to explore the actual properties of these extremals. Generally this is achieved by finding a function space setting for the variational problem and utilizing the solution diamond diagram idea of Appendix 3 of Chapter 1. In the case of Einstein-Kahler metrics discussed above, the variational principles in the various cases were discovered in the last Section, based on the ideas of gradient operator and nonlinear partial differential equations. In this last Section of Chapter 3 , we now explore the function space setting for this problem in the context of Sobolev spaces on compact complex manifolds. As in the Yamabe problem, the boundedness properties of certain Sobolev imbeddings hold, but unfortunately the compactness associated the various Kondrachev results fail; the so-called "limit case "of Sobolev inequalities . To date, this difficulty has never been resolved, so that the discussion here ends with plausible conjectures and the hope that some reader may uncover the mathematical key to this important problem.

Let us denote by I any of the functional $I_+, I_-,$ and I_0 defined in (14). For investigating critical points of I by means of functional analysis, it is natural to work on Sobolev spaces. The appropriate one turns out to be the Sobolev space $W_{2,m}(X)$, defined as the completion of $C^2(X)$ with respect to the norm

$$\|u\|_{W_{2m}} = \|\nabla^2 u\|_m + \|\nabla u\|_m + \|u\|_m$$

(where $|\nabla^2 u|^2 = \frac{1}{2} \nabla^{ab} u \nabla_{ab} u = \nabla^{\alpha\beta} u \nabla_{\alpha\beta} u + \nabla^{\alpha\bar{\beta}} u \nabla_{\alpha\bar{\beta}} u$ and $\| \ \|_p$ is the $L_p(X)$ norm for $p \geq 1$).

Since the real dimension of X is $2m$, we are in the limit case of the Sobolev inclusion theorem and, for any $u \in W_{2,m}(X)$, e^u is integrable with convenient inequalities and compactness properties of the mapping

$$u \in W_{2,m}(X) \to e^u \in L_q(X).$$

More precisely, there exist real numbers A and B such that, if $u \in W_{2,m}(X)$,

(15)
$$\int_X e^u \, dV \leq A \exp \left[B\|\Delta u\|_m^m + \frac{\int_X u \, dV}{\int_X dV} \right]$$

and thus, for some constants C and μ independent of u,

(16)
$$\int_X e^u \, dV \leq C \exp (\mu \|\mu\|_{W_{2m}^m}).$$

On the other hand, the following compactness result holds: if (u_i) is a bounded sequence of $W_{2,m}(X)$, there exist a subsequence, also relabeled (u_i) and an element $u \in W_{2,m}(X)$ such that, for any $q \geq 1$, $\exp(u_i)$ converges strongly to e^u in $L_q(X)$; by reflexivity of

$W_{2,m}(X)$, we can also suppose (u_i) converges weakly to u in $W_{2,m}(X)$.

Proposition 1 extends J and j to $W_{2,m}(X)$ and defines the weak solutions of the deformation equations as the critical points of the extended functional j, the C^2 ones being the admissible C^α solutions of these equations. In Proposition 2, positivity and convexity properties of J are studied.

Proposition 1

(i) H_k (see (12)) can be extended as a continuous (k+1) linear symmetric form on $W_{2,m}(X)$; thus I and J, defined in (18), (17), extend as polynomial C^∞ functionals on $W_{2,m}(X)$.

(ii) j extends as a C^∞ bounded real valued functional on $W_{2,m}(X)$. The zeros of the Fréchet derivative

$$j' \colon W_{2,m}(X) \to W_{2,m}(X)^*$$

are called the weak solutions of the deformation equations. The weak solutions of class C^2 are in fact smooth C^∞ admissible solutions of these equations. We now state the basic geometric properties of the functionals defined in (14) and (14') relative to the Chern class C_1 of prescribed sign for the Kähler manifold (X, g) of dimension 2m.

(i) Case $C_1(X) < 0$ on Π the functional defined in (14) with $\varepsilon = -1$

$$I_-(\phi) = J(\phi) - \int_X \phi \, dV + \int_X e^{\phi + f} \, dV$$

is bounded from below, convex and thus lower semicontinuous with respect to weak convergence.

(ii) Case $C_1(X) = 0$ on Π the functional defined in (14')

$$I_0(\phi) = J(\phi) + \int_X (e^f - 1)\,\phi\,dV.$$

is convex, weakly lower semicontinuous, and bounded from below if

$$\int_X e^f dV = V$$

(iii) Case $C_1(X) > 0$ on Π the functional defined in (14) with $\varepsilon = 1$

$$I_+(\phi) = J(\phi) - \int_X \phi\,dV - \int_X e^{-\phi+f}\,dV$$

is lower semicontinuous with respect to weak convergence on Π. Moreover, on $W_{2,m}(X)$ this functional has the natural constraint

$$\Sigma = \{\, u \in W_{2,m}(X);\ G(u) = \int_X e^{-u+f}\,dV = V \,\}$$

We shall not prove these results here but rather refer the reader to the research paper of Berger and Cherrier listed in the Bibliography at the end of this book. The chief missing component in the associated argument via the direct method of the calculus of variations is the coerciveness of the associated functionals. For example, in the simplest case when the first Chern class is negative: Does $I_-(u)$ tend to infinity, when $\|u\|_{2,m}$ tends to infinity? Here are some conjectures concerning the approach outlined above:

(A) In each of the cases described above, we find a minimizing sequence of the associated functional with (possibly) added constraints. The resulting minimizing sequence has uniformly bounded Sobolev $W_{2,m}$ norms, and thus a weakly convergent subsequence with weak limit U, say. Clearly U is an element of the Sobolev space $W_{2,m}$ and we conjecture U is a critical point of the associated variational problem in each case.

(B) Secondly, we conjecture that this weak limit U, after an appropriate redefinition on a set of measure zero, is smooth enough to generate a positive definite metric on M, as well as the desired Kähler deformation.

Chapter 4   Vortices in Ideal Fluids

Section 4.1   The Early History of Vortices in Fluids

The observation that vorticity is of great importance in nature can be traced back hundreds of years, beginning at least to Leonardo da Vinci and Descartes. For any three-dimensional vector field   v we know the fundamental fact that that provided the vector field v vanishes outside a compact set in R^3, then the vector field v decomposes into a solenoidal part and a gradient part. This fact is sometimes called "the fundamental theorem of advanced calculus" The solenoidal part of v, sometimes denoted   curl v, is called vorticity when the vector field in question is the velocity vector of a fluid. This means, in usual discussions, assuming the velocity is zero at infinity, that the vorticity can be distinguished from the irrotational part of the fluid by a rotational movement. In symbols we have

$\underline{v}$ = velocity vector of a fluid in $\mathfrak{R}^3$, of compact support

curl $\underline{v}$ = vorticity of the fluid in $\mathfrak{R}^3$

grad $\underline{v}$ = irrotational part of $\underline{v}$

$\underline{v}$ = curl $\underline{v}$ + grad $\underline{v}$

On the scientific side, Leonardo da Vinci was the first scholar to observe vorticity carefully and to incorporate it into his magnificent drawings of fluids. See the drawings on the following pages. However, due to the limited mathematical knowledge   of his epoch, Leonardo's ability to use the advanced mathematical ideas neccessary for fluid dynamics, was extremely limited, both by his education and the fact that mathematics, in the Renaissance world in which he lived, was not enough developed to cope with the nonlinear science of vorticity.

A century later, and quite independently, Descartes made vortices the fundamental building blocks of his physical universe. Before Newton, he derived ideas of mechanics which were theoretically attached to vortices and the mechanical consequences of their appearance in nature. His ideas, however, are based on the ubiquity of the so-called "ether," a fluid without viscosity that inhabited the universe. Later physical experiment negated the existence of this ether. Thus, subsequently, Descartes' ideas have taken a subsidiary role, until, of course, the discovery of nonlinear quantum fields.

Nonetheless, great mathematical scientists of the nineteenth century such as Kelvin, took up this idea of vortices as the building blocks of nature. Together with Tate, Kelvin and Tate developed the theory of knots in the geometric sense to describe "vortex atoms." These ideas also took a subsidiary role with the nonexistence of the ether. Nonethelesss, the associated topological ideas were the origin of the mathematical theory of knots.

The first mathematical scientist to put vortices on a firm footing was the great nineteenth mathematical physicist and man of learning, H. Helmholtz. Helmholtz formulated the physical laws of vortices in mathematical terms, applicable to an ideal fluid and formulated various conjectures about vortices that have been of interest up to the present time.

In the sequel, we shall discuss some of Helmholtz's ideas and conjectures about fluids. Moreover, we shall discuss some of their implications in the contemporary science and technology of fluids, as well as some relevant aspects of the problem of vortex motion using the mathematical ideas of nonlinear science mentioned in this book. Other ideas, to be discussed on another occasion, involve meteorology, airplane flight, and oceanography. Indeed, an adequate mathematical treatment of these fields alone, would require another large treatise.

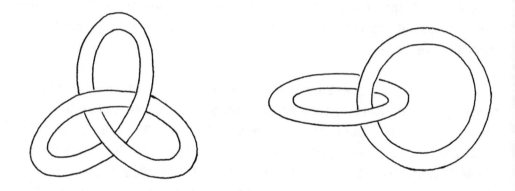

Kelvin's linked vortex rings.

Section 4.2   Formulation of the Vortex Concept in Ideal Incompressible Fluids

The simplest way to begin the discussion of vortex motion in fluids is to make several abstract definitions about the fluid involved. The simplest abstraction involves the notion of an incompressible fluid without dissipation (such fluids are called ideal,while the dissipation is called viscosity) and the motion of such fluids subject to given initial conditions. We begin by denoting the velocity vector of such a fluid by $\underline{v}$. Then the standard fact that the fluid is incompressible in three dimensions is equivalent to saying that the divergence div $\underline{v}$ = 0. To isolate the notion of vortex in such a fluid we use the concept of curl of $\underline{v}$ and we define the vorticity vector of the fluid $\underline{\omega}$ = curl $\underline{v}$.

The simplest kinds of fluid motion in an incompressible fluid have curl $\underline{v}$ = 0 , i.e. the vorticity vanishes throughout the incompressible fluid. Such fluid motions are called irrotational. It was Helmholtz who observed that it is vitally important to consider cases in which the curl of $\underline{v}$ $\neq$ 0 in certain regions in a fluid. One then wishes to determine exactly what the laws of motion for a fluid with nonzero vorticity would be.

For an incompressible ideal fluid the equation of motion was determined by Euler in the eighteenth century. It is, in fact, a consequence of Newton's laws of motion. This equation is known as Euler's equation and is one of the first partial differential equations, arising in the study of nature ,which is nonlinear. This equation can be written in terms of the velocity vector v of the fluid as follows:

(1)
$$\frac{\partial v}{\partial t} - v \times \omega = F - \nabla\left(\frac{p}{\rho} + \frac{1}{2}q^2\right)$$

Here q^2 denotes $|v|^2$, F denotes the given external force and p denotes pressure.

Using the vector identity

$$(v \; \text{grad})v = \text{grad} \; \frac{1}{2}v^2 - v \times \text{curl} \; v$$

and setting the density ρ to be unity , we rewrite (1)

(1')
$$\frac{\partial v}{\partial t} + (v \cdot \text{grad})v = \text{grad} \; p'$$

where in terms of Cartesian coordinates

$$v \cdot \text{grad} = v_1\frac{\partial}{\partial x_1} + v_2\frac{\partial}{\partial x_2} + v_3\frac{\partial}{\partial x}$$

and the external force F, assumed to be the gradient of a function, has been incorporated into the modified pressure p'.

Dynamical equations, of infinite dimensions, such as Euler's equation written above (equation 1 or 1'), are of special interest in this book because they are nonlinear in terms of the unknown velocity vector v. Thus the dynamics predicted by such an equation possess special features that make this topic of great current interest. For example, this Euler equation is not one of the integrable equations discussed in Chapter I. Nonetheless, it possesses a number of features that link it closely with integrable systems of nonlinear partial differential equations. This equation possesses a "soliton" solution, for example. This means a solution that depends on time but nonetheless has certain invariance properties, namely, the "soliton" solution describes a fluid motion which moves with constant speed and permanent shape.

Such solutions, here called "vortex rings," because of their toroidal form, and are observed widely in incompressible fluids. Such a solution was discussed by Helmholtz in his fundamental first research on this topic. More precisely, by a vortex ring we mean, a figure of revolution, (expected to be homeomorphic to a solid torus in most cases) associated with a continuous axisymmetric solenoidal vector field q (the fluid velocity), having the following properties when we take axes fixed in the ring : (a) both the figure of revolution and the vector field q do not vary with time, (b) the vorticity curl q has a positive magnitude within the figure of revolution, vanishes outside it and satisfies a nonlinear equation of motion which, among other things, determines the vortex core, (i.e. the boundary of the figure of revolution and finally (c) q has either a prescribed normal component on the boundary of the domain of definition of the fluid velocity or tends to a constant value at infinity

Helmholtz discovered special dynamical features of the vorticity of the velocity vector field v associated with Euler's equations. These special conservation laws are quite remarkable. To begin, we find the "vorticity equation" associated with an ideal incompressible fluid. One simply takes the "curl" of both sides of (1), recalling ω = curl v. Using vector analysis, we find

$$\frac{\partial \omega}{\partial t} = \text{curl} (v \times \omega)$$

(2)
$$= -v \cdot \text{grad } \omega + \omega \cdot \text{grad } v$$

Here we have used the facts that div $v = 0$ and div $\omega = 0$ and the external force F is the gradient of some potential function.

We now define

$$\frac{\partial \omega}{\partial t} + v \cdot \text{grad } \omega = \frac{D\omega}{Dt} \quad \text{the material rate of change of } \omega$$

Thus (2) can be rewritten as

(2')
$$\frac{D\omega}{Dt} = \omega \cdot \text{grad } v$$

Thus one conclusion from (2') is that the material rate of change of ω is zero wherever ω is zero. Thus we find, with Helmholtz, that fluid particles which were vortex-free at some time remain so forever. The equation (2) also leads to the conservation properties of vortices in an ideal incompressible fluid, which can be paraphrased by saying that in an inviscid fluid of uniform density, a vortex-tube moves with the fluid and its strength remains constant. Thus, in such a fluid, vortices cannot be created or destroyed without the fluid becoming non-ideal somewhere.

An immediate consequence of the invariance of vortex strength in an ideal fluid is that a vortex-tube can never terminate within the fluid. Thus it either reaches the boundary or must determine a closed figure. One can also consider the case when the vorticity is confined to a three-dimensional region with small cross-section q, namely, a vortex filament. In this case, the strength μ of the vorticity remains constant even though $q \to 0$; in fact, we write

$$\mu = \lim_{q \to 0} \omega \cdot q$$

It is quite amazing that recently new conserved vortex invariants for ideal incompressible fluids have been found. For example, K. Moffatt and J. Moreau independently isolated an invariant, now known as helicity H, measuring the degree of knottedness of tangled vortex lines in a blob of confined vorticity in an ideal fluid. This helicity, in the notation defined above, is given by the integral

$$H = \int v \cdot \omega dV$$

The simplest way to continue the investigation of the vorticity of an ideal fluid (in terms of nonlinear science) is to introduce the notion of a vector potential ϕ associated with an incompressible fluid satisfying div $\underline{v} = 0$ on R^3. Namely, we set

$$\underline{v} = \text{curl } \phi$$

which is valid by the well-known Poincaré theorem. Then the vorticity

$$\begin{aligned}
\underline{\omega} &= \text{curl } \underline{v} \\
&= \text{curl curl } \phi \quad . \\
&= -\Delta\phi + \text{grad div } \phi
\end{aligned}$$

Note that we choose div $\phi = 0$ by an elementary gauge change. Thus by the laws of vector analysis, we find

(2) $\underline{\omega} = -\Delta\phi$

Given the vorticity vector $\underline{\omega}$ we can thus find the vector potential ϕ and hence the velocity vector $\underline{v}$ by solving the partial differential equation (2) with appropriate boundary conditions.

It is customary to choose $\underline{\omega}$ to be the simplest possible distribution of vorticity, namely, a delta function concentrated at a point x_0. In this case the vector potential ϕ involves a Green's function which, in higher dimensions, as we know, is singular at the point x_0. This gives rise to the simplest possible "vortex soliton" known as a closed "vortex filament" in an axisymmetric coordinate system. In fact, when adjusted for axisymmetric coordinate axes, the resulting figure is known as Helmholtz's circular vortex filament. It is an interesting abstraction arising from the choice of $\underline{\omega}$ as a delta function in a special coordinate system.

However, this abstraction is less useful for computation than one might imagine because the associated vortex ring moves with infinite speed. Nonetheless one might inquire whether it might be possible to choose the vorticity vector $\underline{\omega}$ to be a smoothed out version of this delta function choice. This would produce a finite energy for the associated vortex soliton and thus a computable vortex ring that propogates with finite speed. This smoothed out version of the delta function proved, however, to be a complicated problem in the nineteenth century. This is exactly a concrete instance of the problem of "nonlinear desingularization" described earlier in this book. It amounts to studying a nonlinear partial differential equation with appropriate boundary conditions, determining the free boundary of the resulting vortex ring, i.e. the support of the vorticity . One can also find the behavior of the vortex ring as it propagates at large distances. In fact, it turned out that his classical problem requires the nonlinear analysis techniques discussed in this book for its resolution.

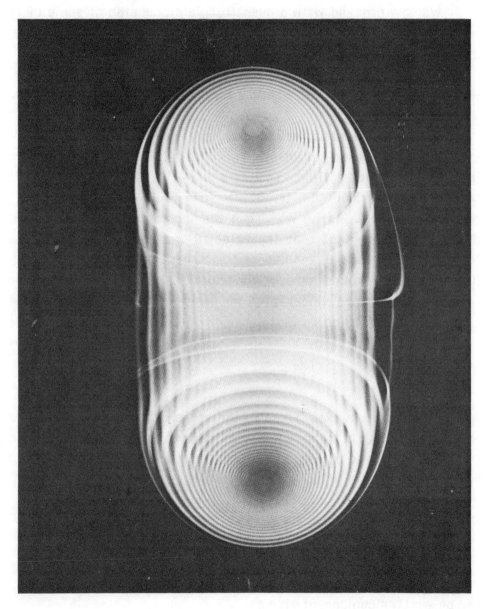

Section 4.3   Axisymmetric Vortex Motions with and without Swirl

We now proceed with a quantitative discussion of some of the points mentioned in the above section. In particular, we shall give a discussion of axisymmetric vortex rings in an ideal fluid.

Governing Equations of Steady Axisymmetric Vortex Flows

For completeness, we derive the governing equations for axisymmetric steady motions of the Euler equation in an incompressible ideal fluid. To this end we let the cylindrical polar coordinates be denoted (r, θ, z) and the associated velocity components $\underline{v} = (u, v, w)$. Then the incompressibility condition div $\underline{v} = 0$ and the fact that the velocity vector does not depend on the angular variable θ enables us to introduce a stream function ψ (known as the Stokes stream function), as a smooth real-valued function of the two variables r and z with the velocity components satisfying

$$ u = -\left(\frac{1}{r}\right)\psi_z \qquad w = \left(\frac{1}{r}\right)\psi_r $$

We take coordinate axes, fixed in the vortex ring to be determined, and demand that, in terms of the cylindrical polar coordinates, and the constant speed of propagation W of the vortex ring, the fluid velocity $v \to (0, 0, -W)$ at infinity.

Then the vorticity vector

$$ \omega = \text{curl } \underline{v} = (\Omega_r, \Omega_\theta, \Omega_z) $$

can be written componentwise as

$$\Omega_r = -\left[\frac{1}{r}\right](rv)_z , \qquad \Omega_\theta = u_x - w_r , \qquad \Omega_z = \left[\frac{1}{r}\right](rv)_r .$$

Consequently

(1)
$$r\Omega_\theta = -(\psi_{zz} + \psi_{rr} - [1/r]\psi_r) = -L\psi$$

where L is the axisymmetric Laplace operator. Here the steady Euler equation can be written (in non-dimensional form) as

(2)
$$\underline{v} \times \underline{\Omega} = \nabla H(x)$$

where assuming the density $\rho = 1$, the enthalphy $H = 1/2 |v|^2 + p$. These three equations in terms of (u, v, w) and $(\Omega_r, \Omega_\theta, \Omega_z)$ can be written (after simplification)

(3a)
$$-\frac{\partial H}{\partial r} + \frac{v}{r}\frac{\partial}{\partial r}(rv) + \frac{1}{r^2}\psi_r (L\psi) = 0$$

(3b)
$$\frac{1}{r^2}(\psi_r(rv)_z - \psi_z(rv)_r) = 0$$

(3c)
$$-\frac{\partial H}{\partial z} + v(v)_z + \left(\frac{1}{r^2}\right)\psi_z (L\psi) = 0$$

The second of these states that the Jacobian

$$J\left(\frac{\psi, rv}{r, z}\right) = 0 \quad \text{identically.}$$

Thus rv is a function of ψ alone, i.e.,

(*) $rv = f(\psi)$ for some function f = the swirl function.

In the traditional texts, the swirl function f vanishes identically; the associated vortex motions are without swirl.

Thus the remaining two equations of (3) can be written

$$- \frac{\partial H}{\partial r} + \frac{\partial \psi}{\partial r} \left(\frac{1}{r^2} L\psi + \frac{1}{r^2} ff' \right) = 0$$

$$- \frac{\partial H}{\partial z} + \frac{\partial \psi}{\partial z} \left(\frac{1}{r^2} L\psi + \frac{1}{r^2} ff' \right) = 0$$

After eliminating $1/r^2(L\psi + ff')$ from these equations we find

$$\frac{\partial H}{\partial r} \frac{\partial \psi}{\partial z} - \frac{\partial H}{\partial z} \frac{\partial \psi}{\partial r} = 0$$

Again this implies the Jacobian

$$J\left(\frac{H, \psi}{r, z} \right) = 0$$

This implies H is a function of ψ alone, i.e., $H = -c(\psi)$ identically. Thus the equations reduce to

(4) $L\psi + ff'(\psi) + r^2 c'(\psi) = 0$

(In equation (4) we allow an ambiguity of sign because we do not specify whether the function c is positive or negative.)   This equation is the desired nonlinear elliptic partial differential equation we have in mind defining the stream function ψ in terms of the swirl and Bernoulli functions. When the swirl vanishes we have the reduced equation

$$(5) \qquad\qquad L\psi + r^2 c'(\psi) = 0$$

to determine ψ in terms of the function c, by an equation that is generally nonlinear in ψ. The equation (4) is called the Squire-Long equation.

Instead of studying vortex rings associated with (4), we study the simpler problem as indicated above, namely, how to establish such rings for (5). Nonetheless, the results we establish for (5) have an exact analogue for (4). In fluid terms, the vortex rings we establish for fluids without swirl are also valid for fluids with swirl, although the mathematical proof is somewhat more elaborate.

There are a few key observations in the mathematical study of the smooth finite solutions of equation (5) connected with vortex rings. First, equation (5) can be regarded as the Euler-Lagrange equation of a functional F relative to a function space that incorporates the relevant boundary conditions. However, the associated critical point representing the solution of (5) is not a minimum of the functional F, but rather a saddle point. To study such a problem, utilizing the methods of this book, we convert (5) into a nonlinear eigenvalue problem, by introducing a parameter λ, in this case representing "vortex strength." Thus we set

$$(6) \qquad\qquad c'(\psi) = \lambda g(\psi)$$

One can then utilize the "solution diamond idea" introduced in the appendix at the end of Chapter 1 to find a critical point of the problem and so find a classical solution of equation (5). If one is careful in formulating the exact physical problem, one can even insure that the classical solution so found represents the actual vortex ring desired.

Known Solutions

Standard examples yielding vortex rings are as follows:

1) **Hill's Vortex** (dating back to a paper of M. J. M. Hill in 1894)

This vortex has no swirl, has flux constant $k = 0$ (see a later discussion for the meaning of this constant), and can be represented as a spherical vortex where the vorticity is confined to the solid sphere of radius a in $\mathbb{R}^3$. Exterior to this sphere, all motion is irrotational. Neighboring solutions are known to be thickened tori. More explicitly, we let the function of a single variable $c(t)$ in the modified equations (5) and (6) be defined as $c(t) = 1$ for $t > 0$, and $c(t) = 0$ for $t \le 0$. Then the cross-section S_H of Hill's spherical vortex in cylindrical polar coordinates in the half-plane π can be written

$$S_H = \left\{ (r, z) \in \pi \mid r^2 + z^2 < a^2 \right\} .$$

Set $\rho^2 = r^2 + z^2$. Then Hill found the solution for (5) and (6) as

$$\psi + \frac{1}{2}Wr^2 = \begin{cases} \frac{1}{2}Wr^2 \left(\frac{5}{2} - \frac{3}{2}\frac{\rho^2}{a^2} \right) & \rho \leq a \\[3mm] \frac{1}{2}Wr^2 \frac{a^3}{\rho^3} & \rho \geq a \end{cases}$$

Here the parameters λ, a and W are related by the formula

$$\lambda a^2/W = 15/2$$

showing that a whole family of spherical vortex solutions, with varying radii, exists for the Euler equations, provided the parameter W is varied.

2) Helmholtz's Circular Vortex Filament

In Helmholtz's original paper a circular vortex ring was found with infinitely thin cross section. Currently such infinitely thin cross sections are represented by a vorticity distribution ω represented as an appropriately chosen delta function. The relevant partial differential equation is

$$L\psi = \psi_{zz} + \psi_{rr} - [1/r]\psi_r = -r\,\omega$$

where the vorticity ω has support at a single point of the half-plane π, and yet the integral of the vorticity over π is a nonzero constant. Denoting the operator on the left of the above equation by L, we note that computations of the vortex ring can be carried out by the following representation for the fundamental solution P of L

$$P(r_0, z_0, r, z) = \frac{r_0 r}{4\pi} \int_{-\pi}^{\pi} \frac{\cos\theta \, d\theta}{\left\{ r^2 + r_0^2 - 2rr_0 \cos\theta + (z - z_0)^2 \right\}^{1/2}}$$

The process of thickening the vortex filament to achieve a regular, nonsingular, toroidal vortex is called desingularization, as has already been mentioned several times in this book . The virtue of this process is that the singular, circular vortex filament has an infinite speed of propagation but the desingularized vortex has a finite and calculable speed of propagation.

3) Vortex Rings of Finite Cross Section

In a 1974 paper mentioned in the bibliography, Berger and Fraenkel found a whole family of smooth, toroidal vortex rings interpolating between the Hill's spherical vortex and Helmholtz' spherical vortex filament. All these vortices have no swirl. It is an unusual fact that, analytically, the problem of vortex rings, with or without swirl, has a similar mathematical structure. In the following paragraphs the key mathematical ideas related to finding these vortex rings of finite cross section will be outlined.

4) Moffatt's Spherical Vortex

Recently, K. Moffatt determined generalizations of Hill's spherical vortex by adding considerations of swirl. Moffatt's vortices are spherical as well and can be explicitly computed in terms of Bessel functions based on formulae for the Laplacean in specialized coordinate systems. Moffatt also calculated the helicity H of the spherical vortices in question. The nonzero helicity, in certain cases, of these flows implies the associated streamlines are torus knots with interesting nontrivial knottedness and linkage properties.

To find these spherical vortices, we consider the equation (4) of Section 4.3 and we chose $c(\psi) = c_0 + \lambda\psi$ and $g(\psi) = \alpha\psi$, where c_0, λ, and α are arbitrary constants. The relevant partial differential equation (4) then becomes linear and can be written

$$L\psi = \lambda r^2 - \alpha^2\psi$$

This equation admits solutions in spherical polar coordinates (R, θ, ϕ). These solutions ψ are given by fractional Bessel functions and can be matched, preserving continuity in the relevant physical quantities, by a suitable choice of constants to an irrotational exterior flow. These solutions ψ have vorticity confined to a solid sphere of radius a, vanish at the boundary of the sphere, and represent velocity distributions that are continuous across the free boundary. The associated spherical vortices also propagate at a constant speed W and, in general, have nonzero helicity.

Section 4.4   Variational Principles for the Stream Function for Vortex Rings without Swirl

We begin by considering steady vortex rings without swirl in a bounded axially symmetric domain D with the property that the vorticity ω of the flow is concentrated in a (strict) subdomain Ω of D. Thus the steady flow is separated (i.e., solenoidal in Ω and irrotational in D − $\overline{\Omega}$). The set ∂Ω, the boundary of Ω, is free, i.e. it is determined by the solution of (4) of Section 4.3. This free boundary problem can be written

$$\psi_{rr} - \frac{1}{r}\psi_r + \psi_{zz} = \begin{cases} -\lambda r^2 f(\psi) & \text{in } \Omega \\ 0 & \text{in } \pi - \overline{\Omega} \end{cases}$$

Notational Change. The function f(ψ) used above was denoted g(ψ) in Section 4.3, whereas f(ψ) was used for vorticity. However, in this section, we reverse the roles of these functions.

To formulate this free boundary problem we must add boundary conditions to the problem. We begin by approximating the half-plane π by the bounded domain π(a, b), which we assume is a rectangle in the (r, z) plane with vertices at (0, ± b), (a, b), (a, − b). It is also convenient to decompose the total stream function ψ by writing

(1) $\psi(r, z) = u(r, z) - \frac{1}{2}Wr^2 - k$

where u is the stream function of the velocity field induced by the vortex ring, and $-\frac{1}{2}Wr^2 - k$ represents the stream function of a uniform flow with velocity $(0, 0, -W)$, and $L(\frac{1}{2} Wr^2 + k) = 0$ where k is the positive flux constant, where $2\pi k$ is the flow rate between the axis of symmetry and the boundary of the vortex ring. Thus, ψ at $r = 0$ satisfies the boundary condition $\psi(0, z) = -k$.

Thus, in terms of the function u, we seek a solution of the problem

(2a)
$$Lu = \lambda r^2 f(\psi) \quad \text{on } \pi(a, b)$$

with the Dirichlet boundary condition

(2b)
$$u\Big|_{\partial\pi(a,b)} = 0$$

Here we assume that the function f is a Hölder continuous function on the real line, is nondecreasing, satisfies a polynomial growth condition at infinity and $f(t) = 0$ for $t \leq 0$ while $f(t) > 0$ for $t > 0$. The maximum principle for the operator L implies that if $f(\psi) > 0$ in Ω, then $\psi > 0$ in Ω, and that $\psi < 0$ in $\pi(a, b) - \overline{\Omega}$. Thus we can define the core of the vortex ring Ω as follows

$$\Omega = \{ (r, z) \mid \psi(r, z) > 0 \}$$

Thus the free boundary $\partial\Omega$ is no longer a source of difficulty for the vortex ring problem, but arises naturally as the set of zeros of the function ψ.

We transform this system (2a, b) to a generalized solution (gradient operator equation) on an appropriate Hilbert space H. Here the space H can be chosen to be the closure of the C_0^∞ equations on the rectangle $\pi(a, b)$ relative to Hilbert space inner product

(3) $(u, v) = \iint\limits_{\pi(a,b)} (\frac{1}{r^2}) (u_r v_r + u_z v_z) r\, dr\, dz$

Notice that the Sobolev inequalites relative to L_p for the domain $\pi(a, b)$ can be modified to hold for this Hilbert space. Moreover, sets bounded in H are relatively compact in this L_p space relative to the volume element rdrdz, as follows from the Kondrachev theorem. These facts are, of course, crucial for the nonlinear eigenvalue problems that we discuss. Notice that for the infinite domain π, the nonlinear eigenvalue problem in question loses compactness, but this can be overcome by an approximation argument on the domains $\pi(a, b)$ by letting a and b tend to infinity. Such an argument requires a priori bounds for the vortex cores in question, but the derivation of these a priori bounds is omitted here.

Relative to this inner product the appropriate generalized solution of the adjusted system (2a, b) can be written conveniently as

(4) $(u, \phi) = \lambda \iint\limits_{\pi(a,b)} f(\psi)\phi\, r\, dr\, dz$ for all $\phi \in H$

Thus the associated isoperimetric variational principle is

(I)   Maximize the functional $J(u) = \iint\limits_{\pi(a,b)} F(\psi)\, r\, dr\, dz$ over

         the sphere $\| u \|_H^2 = R$,   R a positive constant

where $F' = f$ and $r \, dr \, dz = d\sigma$ is the appropriate volume element for cylindrical polar coordinates. We note that the smooth critical points of (I) coincide with the solution of (4) in Section 4.3.

In this formulation of the vortex ring problem, the kinetic energy of the vortex motion is proportional to $\| u \|_H^2$, and the positive constants W and the flux constant k are held fixed, but the vortex strength parameter λ is allowed to vary. The variational problem (I) and its associated Euler-Lagrange equation (4) can thus be regarded as fairly standard nonlinear eigenvalue problems for gradient operators.

It turns out that the nontrivial solution of this nonlinear eigenvalue problem (u, λ) can be easily determined by the general principles of Chapter 2, since the bounded domain $\pi(a, b)$ insures the necessary weak continuity of the functional $J(u)$ on the Hilbert space H. Moreover, the nonlinear eigenfunction $u(r, z)$ can be chosen to be nonnegative in $\pi(a, b)$, and in addition u has the symmetry property $u(r, -z) = u(r, z)$ as can be obtained by Steiner symmetrization. Notice this formulation of the vortex ring problem determines the cross-section of the vortex core in question. The full vortex core is obtained by rotating the cross-section about the z-axis.

The result we obtain can be summarized as follows.

Theorem Given the stream velocity $W > 0$, the positive flux constant k, the kinetic energy of the vortex motion, and the vorticity function f satisfying the conditions stated above, a smooth vortex stream function u satisfying the nonlinear eigenvalue problem (2a, 2b) can be found. This stream function satisfies the isoperimetric variational characterization (I) mentioned above.

Corollary There is a family of axisymmetric vortex rings for an incompressible inviscid ideal fluid satisfying the conditions of the above theorem. The vortex rings in question interpolate between Hill's spherical vortex and Helmholtz's circular vortex filament.

The proof of these results just stated for vortex rings in an unbounded domain require two extensions of the results obtained on $\pi(a, b)$, first to the full unbounded domain π and secondly to jump discontinuities in f to cover vorticity distributions similar to Hill's spherical vortex. Both of these extensions can be obtained from this abstract result by fairly standard approximation arguments. The calculus of variations formulation of the vortex ring problem has many other advantages. For example, if the function f is convex, it is fairly easy to show that the core of the free boundary is simply connected. In addition, the calculus of variations formulation also can be extended to cover vortex rings with swirl, since the classical solution of equation (4) of Section 4.3 can be transformed into a gradient operator equation in the appropriate Hilbert space H. Thus, a variational principle for vortex rings with swirl is easily obtained.

Section 4.5   Leapfrogging of Vortices

After one determines the analog of a vortex soliton, namely, in our case a vortex ring, one wishes to describe the interaction between two such solitons. In the classic case of the KdV equation described in Chapter 1, the unusual property of these one-dimensional solitons was that after a time the interactions were not affected by collisions, except for a small change in the phase of the solitons. However, in the case of vortices, the interactions of vortex solitons associated with Euler's equations are quite different. Helmholtz predicted on physical grounds the notion of "leapfrogging" of vortices, i.e. that two vortex solitons would interact in such a way as the motion of the two solitons would produce a relative motion that is periodic. Namely, two coaxial vortex rings would behave as follows: The trailing vortex would catch up and pass through the leading vortex ring. Thus the trailing vortex ring, after the interaction, would become the leading vortex. Then, the vortex ring that was originally the leading vortex ring would then become the trailing vortex after the interaction and this vortex in turn would pass through the vortex ring that had surpassed it. This relative motion would then repeat in a periodic way. This gives rise to the notion of leapfrogging.

The key question here concerns whether this hypothetical situation is actually predicted by the Euler equations of motion, or whether this idea of leapfrogging is simply an idle speculation. In the sequel, we shall show that we can use the notion of vortex filament to exhibit the phenomenon of leapfrogging of vortices. This situation turns out to be somewhat singular in the mathematical sense, as we shall see below.

The leapfrogging of vortex rings in an ideal fluid in three dimensions was first described by Helmholtz in his famous paper on vortex motion. To describe this interaction of two infinitely thin vortex rings of equal strength propagating along a common axis, Love, in 1893, found a qualitative analysis by assuming the two thin vortex rings could be represented by two pairs of two point vortices of equal strength placed symmetrically along a common axis. Leapfrogging in this case means that the relative motion of the two pairs of point vortices is periodic. Recently, Aref showed that the motion of 4 point vortices could be chaotic if the point vortices are not placed symmetrically. Here we demonstrate the leapfrogging phenomenon for two pairs of point vortices of different strengths but symmetrically placed, using the techniques of this book. This result is due to my former student, J. Nee.

We analyze this problem as a bifurcation process from an equilibrium for a Hamiltonian system. The usual approach of analyzing the geometry of level surfaces of the Hamiltonian seems difficult in this case.

§1. The Physical Problem

Consider two pairs of two point vortices propagating along the x-axis (as common axis) initially symmetrically placed in $\mathfrak{R}^2$ with Cartesian coordinates $P_1 = (x_1, y_1)$, $P_1' = (x_1, -y_1)$, $P_2 = (x_2, y_2)$ and $P_2' = (x_2, -y_2)$. These point vortices in $\mathfrak{R}^2$ are to represent two pairs of cylindrical vortices of infinitely thin cross-section of different strengths. Here the circulation about the two point vortices of each pair are chosen equal and of opposite sign. The line of symmetry coincide for each pair (see Figure 1). Thus we choose the strength of $P_1 = \omega_1$, $P_1' = -\omega_1$, $P_2 = \omega_2$ and $P_2' = -\omega_2$. We set the ratio of vortex strengths $\frac{\omega_1}{\omega_2} = \lambda$.   The case $\lambda = 1$ is classical   as mentioned above. We show that if λ is any number between 1 and λ_c (a critical number approximately equal to .61623), the same leapfrogging phenomenon occurs as in the classic case.

In general for n point vortices with coordinates (x_i, y_i) ($i = 1, 2, \ldots n$) it is known that the motion of these n point vortices is governed by a Hamiltonian system

(1) $\dot{z} = J\nabla H(z)$

where $\underline{z} = (\underline{x}, y)$ and J is the matrix $\begin{bmatrix} 0 & -1 \\ 1 & 0 \end{bmatrix}$ and

$$H(x_i, y_i) = \sum_{i \neq j} \omega_i \omega_j \log\left[(x_i - x_j)^2 + (y_i - y_j)^2 \right].$$

Thus, in the special case of the 4 symmetrically placed point vortices with strength $\alpha_1, -\alpha_1, \alpha_2, -\alpha_2$ we first find the conservation law

$$\alpha_1 y_1 + \alpha_2 y_2 = c \qquad c = \text{constant}$$

Consequently, the Hamiltonian H simplifies to (in this case)

$$H = \alpha_1 \alpha_2 \log \frac{(x_1 - x_2)^2 + (y_1 + y_2)^2}{(x_1 - x_2)^2 + (y_1 - y_2)^2}$$

$$+ 2\alpha_1^2 \log 2y_1 + 2\alpha_2^2 \log 2y_2$$

We have supposed above that $\alpha_2 = \lambda\alpha_1$. Indeed, without loss of generality we suppose $\alpha_2 = 1$. Thus, to study relative motion we set

$$x = x_1 - x_2 \qquad y = y_1 - y_2$$

Thus the relative motion is described by the Hamiltonian system

(2) $$\dot{x} = H_y(x, y, \lambda)$$

$$\dot{y} = - H_x(x, y, \lambda)$$

$$\text{with } H = (1+\lambda)\left[\log\left[x^2+(f+gy)^2\right]\left[x^2+y^2\right]^{-1} + \frac{1}{\lambda}\log(c+\lambda y) + \lambda \log(c-y)\right]$$

$$f = 2c(\lambda+1)^{-1} \qquad \text{and} \qquad g = (\lambda-1)(\lambda+1)^{-1}$$

where the Hamiltonian H above can be simplified to depend only on x, y and λ and the constant c defined above. In the sequel we shall demonstrate the leapfrogging of the two vortex pairs $P_1 P_1'$ and $P_2 P_2'$ on Figure 1 by proving

Theorem   The system (2) has a smooth non-constant periodic solution for each fixed $\lambda \in (\lambda_0, 1)$, i.e. the relative motion of the vortex pairs is periodic. Here λ_0 is a critical parameter described in Section 2, and is approximately equal to .61623.

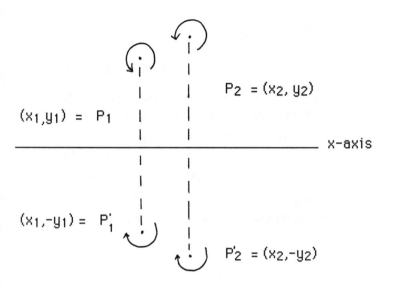

$(x_1, y_1) = P_1$

$P_2 = (x_2, y_2)$

x-axis

$(x_1, -y_1) = P_1'$

$P_2' = (x_2, -y_2)$

FIGURE 1:   Coordinates for the two pairs of 2 point vortices

§2. Analysis of the Hamiltonian System

We analyze the Hamiltonian system (2) by first showing that for λ in a large open interval $(\lambda_0, 1)$ that the system (2) has a unique equilibrium point $(0, y_\lambda)$ where y_λ is a real number implicitly defined by the unique real root of a cubic equation with two other complex conjugate roots. This analysis is carried out in an Appendix.

Then one writes the Hamiltonian system (2) in the form

(3) $\dot{z} = \dot{\ }L(\lambda)z + O(|z|^2, \lambda)$

for $L(\lambda) = H_{z\bar{z}}(0, y_\lambda)$ the Hessian of the Hamiltonian, near the singular point.   One proceeds to show that the desired periodic solution of (3) bifurcates from this stationary point for each λ in the open interval $(\lambda_0, 1)$. To achieve this result we proceed by introducing a new parameter ω into (3) by changing the time scale $t = \omega s$. This allows the study of solutions of period ω in t to be reduced to the study of   1-periodic solutions in   s.   The equation (3) becomes the nonlinear eigenvalue problem

(4) $\dfrac{dz}{ds} = \omega\left\{ L(\lambda)z + O\left(|z|^2, \lambda\right) \right\}$

To analyze the 1-periodic solutions of   (4)   it will suffice to consider the linearized problem about the stationary point $(0, y_\lambda)$

(4') $\dfrac{dZ}{ds} = \omega L(\lambda)Z$

§3.   Analysis of the linearized problem   (4')   and the nonlinear problem (4)

We consider now the periodic solutions $Z(s)$ of (4') in terms of (x, y) coordinates. Note first that $H_{xy}(0, y_\lambda) = 0$. Thus (4') simplifies to

(5)
$$\frac{d\tilde{x}}{ds} = \omega a(\lambda)\tilde{y}$$

$$\frac{d\tilde{y}}{ds} = -\omega b(\lambda)\tilde{x}$$

where a and b are constants depending on the parameter λ . Clearly we have periodic solutions for positive ω provided the product ab is positive. Indeed both $\tilde{x}(s)$ and $\tilde{y}(s)$ satisfy the second order equation

$$v_{ss} + \omega^2 a(\lambda)b(\lambda)v = 0$$

So $\tilde{x}(s)$ and $\tilde{y}(s)$ will be periodic solutions provided the product $a(\lambda)b(\lambda)$ is positive. Using numerical analysis we find that the product $a(\lambda)b(\lambda) > 0$ provided λ is in the open interval (.61626, 1). However, for $\lambda = \frac{1}{2}$ $a(\lambda)b(\lambda) < 0$, so if λ is outside this open interval, the result fails.

The nonlinear result will follow immediately from the standard Liapunov-Schmidt method for bifurcation theory, once one restricts attention to periodic solutions of (4) and (4') that are odd in x(s) and even in y(s). Here we use the symmetry of H in (2), H(x, y) = H(-x, y). This yields the fact 1-periodic solutions of (4') with the stated symmetry is one dimensional, a required prerequisite for the applicability of the standard theory. In order to utilize the Liapunov-Schmidt technique we simply note that there is a mapping A defined by (4) between the Hilbert spaces of periodic functions X and Y with

$$X = \left\{ \begin{array}{l} (x, y) \text{ of 1-periodic functions in } W_{1,2}\left[(0,1),\mathcal{R}^2\right] \text{ with } x(t) \text{ odd} \\ \qquad\qquad\qquad\qquad\qquad\qquad\qquad\qquad y(t) \text{ even} \end{array} \right\}$$

$$Y = \left\{ \begin{array}{l} (x, y) \text{ of 1-periodic functions in } L_2\left[(0,1),\mathcal{R}^2\right] \text{ with } x(t) \text{ even} \\ \qquad\qquad\qquad\qquad\qquad\qquad\qquad\qquad y(t) \text{ odd} \end{array} \right\}$$

To this mapping A the Liapunov-Schmidt method is applicable. Indeed $A(z) \equiv \dfrac{dz}{ds} - \omega\left[L(\lambda)z + 0\left(|z|^2, \lambda\right)\right]$ is a C' Fredholm mapping of index zero between the Banach spaces X and Y that can be used for the standard Liapunov-Schmidt theory.

The theory results in a nonzero-periodic solution $z_\omega(s)$ bifurcating from the equilibrium state $(0, y_\lambda)$ of (4) for each λ in the open interval $(\lambda_0, 1)$. In fact these periodic solutions correspond to the nonconstant ω-periodic solutions of (2) as stated in the Theorem of §1.

Appendix

We write the Hamiltonian system (2) of §1 in the form

$$\dot{x} = h_1(x, y, \lambda)$$

$$\dot{y} = h_2(x, y, \lambda)$$

We prove

Lemma   The equilibrium point of the system (2) is uniquely determined by the parameter λ in the open interval $(\lambda_0, 1)$.

Proof   To determine the equilibrium points of (2) we note that

$$h_2(x, y, \lambda) = x\, c_\lambda(y)$$

where the smooth function $c_\lambda(y)$ vanishes at points where $h_1 \neq 0$. Thus the equilibrium points of (2) must be of the form $(0, y_\lambda)$ if they exist at all. We then determine the desired equilibrium points by solving

$$h_1(0, y, \lambda) = 0$$

Here y satisfies the equation

$$y^3 + \frac{3(\lambda-1)c}{\lambda^2 + 1}\, y^2 - \frac{6c^2}{\lambda^2 + 1}\, y + \frac{4c^3}{(\lambda^2+1)(\lambda-1)} = 0$$

Set $y_0 = cA$, then after simplifying the number A satisfies a cubic equation with coefficients depending only on λ. This equation has a unique real root A for each $\lambda \in (\lambda_0, 1)$.

Section 4.6   Vortex Breakdown

Overview

Vortex breakdown involves the fact that, under certain conditions, a vortex, possessing axial flow and swirl, in an incompressible fluid exhibits a sudden transition from one state to another quite different flow. Such phenomena occur in such diverse fields as aerodynamics and combustion. In addition, breakdown is observed to occur in different types, bubble (meaning axisymmetric) and spiral (meaning nonaxisymmetric with a spiral symmetry) breakdown being the most common. In this section we discuss the onset of vortex breakdown based on a steady inviscid model. We first consider the governing Euler equations in a stream function formulation with its parameter dependence, the so-called Squire-Long equation mentioned in Section 4.3. We demonstrate here some mathematical methods for obtaining multiple solutions to simplified forms of this equation subject to appropriate boundary conditions. The resulting coherent nonlinear vortex structures can be understood as a bifurcation phenomenon involving separated flows. The results obtained are related to Benjamin's subcritical-supercritical vortex-breakdown theory (terms to be defined later) and to recent computer studies involving bubble-type vortex breakdown and axial symmetry.

Introduction

Vortex breakdown has been discussed by numerous authors since the early 1960's when it was observed experimentally in aircraft wing design.   Since then its understanding and predictability have been the subject of a heated debate among experimentalists and theoreticians. It is the purpose of this section to explore some modern nonlinear mathematical viewpoints concerning this issue. Here it is remarkable, we believe, that new mathematical developments can add new points of view that could not be obtained by physical experiment alone. The reason for this situation is clear: the issue is highly nonlinear, so linear intuitions and the normal linear structures used to discuss fluid instabilities and the like must be supplemented by modern mathematical, analytic and computational efforts.

Historically, a major advance in the theoretical developments of vortex breakdown was made by H. B. Squire (1960). This was further developed by T. B. Benjamin in a well-known article (1962) listed in the Bibliography. It was Benjamin who pointed out that vortex breakdown could be studied as a finite transition between two conjugate states of inviscid fluid flow. This analysis was significant because it yielded predictions that could be checked against experiment. Previous studies of vortex motion omitted a coherent discussion of swirl, a topic important in vortex breakdown studies. Nonetheless, the tools developed in the study of vortex rings described above turn out to be easily adapted to this more physical problem .

A well-known photograph shows two types of vortex breakdown that are fundamental for the understanding of the breakdown mechanism. The upper half shows the so-called "bubble" type of vortex breakdown. The lower half of the picture shows the "spiral" type. These vortices are shed by the sides of a triangular wing immersed in water. In the sequel we shall indicate that the different types of breakdown can be discussed from the point of view of symmetry of the nonlinear structures involved. We shall begin our discussion with the quantitative description below and continue with the breakdown mechanism, illustrating these mechanisms with the relevant mathematical and analytical structures.

Governing Equations

Here we shall discuss the relevant quantitative fluid dynamics involved in the vortex breakdown problem. We begin with assumptions on the type of fluid involved in the vortex phenomena.

First approximation It is traditional to consider the onset of vortex breakdown in a fluid that is incompressible and inviscid, thus the Euler equations of motion of an ideal fluid are involved. It is also traditional to begin a discussion with the assumption of steady motion. Since even this physical system cannot be analytically treated in depth at the moment, we make an additional assumption of symmetry for the physical problem. In particular, we shall mention two kinds of symmetry, first axial symmetry in which we consider a velocity vector field for the fluid depending only on the radial and axial coordinates in symbols $\underline{v} = \underline{v}(r, z)$. This symmetry together with the assumption of axisymmetry enables us to introduce the Stokes stream function u for the problem and reduces the mathematical problem from a complicated system to a single equation. The relevant equation of motion in this context is called the Squire-Long equation. It can be written (cf. equation (4) of Section 4.3) in terms of the Stokes stream function $u = u(r, z)$ in cylindrical polar coordinates

$$(1) \qquad u_{rr} - \frac{1}{r} u_r + u_{zz} = r^2 H'(u) - I'(u)$$

plus appropriate boundary conditions. Here the functions H(u) and I(u) have physical meaning. H is known as the head function and I is apart from a numerical factor, equal to the square of the circulation G. The fact that the total head function H and the circulation G depend only on u reflects the conservation of energy and circulation for the ideal fluid.

This equation is also important because of the parameter dependence that it displays when the relevant physical properties are scaled to be nondimensional. If the radial and axial variables are made nondimensional and if the circulation G and the stream function u are scaled accordingly, we find that the Squire-Long equation involves only one parameter in the present context. We shall call this the Squire parameter S. It represents the ratio of an axial velocity to a wave velocity and is introduced most easily into equation (1) by putting an S^2 parameter as a coefficient for the I'(u) term.   This equation can be interpreted as a nonlinear eigenvalue problem with the parameter S^2 as the nonlinear eigenvalue. In the sequel, we shall use other parameters in place of S^2, but their physical meaning is the same. Problems associated with equations like (1) have interesting solution structures with new secondary solutions appearing when the parameter changes. We will find in the sequel that it is exactly this mathematical situation that gives the first clue as to the behavior of vortex structures in this context.

Second Approximation If viscous effects are taken into account, the Euler equation formulation mentioned above must be supplemented and attention focussed on the relevant Navier-Stokes equation-vortex formulation. A number of recent analytical studies are devoted to this issue. Special results are obtained by replacing the analytic results by computation and computer graphics using modern high speed computers. We defer a discussion of this approximation to a later occasion.

Remarks on Nonlinear Dynamics

The subject of vortex breakdown is a part of the major mathematical area known as nonlinear dynamics. This area is going through a major upheaval at the moment due to the discovery of chaotic behavior in many well-known dynamical models with large dissipation.

Relative to vortex dynamics, it is rather clear that viscous effects are somewhat negligible for the physical problem involved. Thus, the relevant dynamical system has no dissipation in the first approximation. Hence, the relevant dynamical system in the first approximation is a conservative one. The major qualitative behavior is due to a change in the relevant physical parameter, namely, the Squire parameter denoted hereafter by S. Turbulence is not a consideration here, at least at the onset of the breakdown phenomena.

Thus, the inviscid vortex dynamics considered in this section can be compared with the classic dynamical system known as billiards. There is a major caveat here however; normally billiards are played on a flat surface. The interesting nonlinear fact that has to be faced for inviscid vortex dynamics in this problem is that this flat surface is changed to a playing field of infinite dimension consisting of mountain ranges with large peaks and valleys.

Moreover, in this section, we are studying inviscid vortex statics for the onset of vortex breakdown. The analogy here is that we are studying one particular mountain range mentioned above carefully. This approach leads us to be able to avoid chaotic behavior and at the same time to be able to bring advanced parts of mathematical analysis and computation to this important problem.

The Viewpoint

Here we ask the question, "What are the non-perturbative nonlinear structures hidden in the physical problem of vortex breakdown?" To answer this question we shall first look carefully at the governing equation for the static problem, namely the Squire-Long equation, and determine the effect of large swirl on the mathematical problem involved. This translates into ascertaining behavior of the static problem for large Squire parameter and large amplitude stream functions.

In one space dimension for problems which do not involve the axial component of velocity, Benjamin, in the paper cited above, rewrites the dynamical problem in terms of the Euler-Lagrange equations of a flow-force Lagrangian S given by the formula for a bounded cylindrical flow with $y = \frac{1}{2} r^2$ and y varying between 0 and a

$$S(\psi) = \text{const.} \int_0^a \left\{ \frac{1}{2} \psi_y^2 + H(\psi) - \frac{1}{2y} I(\psi) \right\} dy .$$

Benjamin analyzed the following examples of the mathematical vortex breakdown problem in terms of a transition from supercritical to subcritical solutions of the reduced Squire-Long equation. Here a supercritical flow ψ means that $S(\psi)$ is an absolute minimum over an appropriate class X of comparison flows and a subcritical flow ψ means that $S(\psi)$ is a saddle point relative to the comparison class X.

Example 1

The following example was first discussed in the context of vortex breakdown by H. B. Squire and then developed by T. B. Benjamin. The example consists of linear equations of motion with the free boundary governed by continuity considerations thus yielding a nonlinear problem. In the example we shall find a supercritical primary flow occurring when a parameter is small. When the parameter increases beyond a certain critical limit this primary flow turns subcritical. A subcritical secondary flow, called B, also occurs here. The subcritical B conjugate flow exists when the parameter is sufficiently large. It is a small perturbation of the primary flow for the parameter near critical limit.

The primary flow consists of two parts: a core with solid body rotation in part one, surrounded by an annular region of irrotational fluid motion. The free boundary for the primary flow, here called flow A occurs with the radial variable r = 1. This free boundary r = 1, separates the two parts of the flow for the primary case. The secondary flow, here called B, has a free boundary displaced to $r = r^*$, again possessing an irrotational part outside the sphere $r = r^*$ and a rotational part inside. Expressed analytically, we assume a stream function ,depending only on the radial variable r , called u = u(r) exists, so that the problem reduces to an ordinary differential equation for u and all considerations are independent of the axial coordinate z. Thus we choose a primary flow A, written U_A, satisfying the Squire-Long equation with

$$H'(u) = 2\omega^2$$
$$I'(u) = 4\omega^2 u$$

The equation of motion thus becomes

$$Lu = \begin{cases} 2\omega^2 r^2 - 4\omega^2 u & \text{inside} \\ 0 & \text{outside} \end{cases}$$

with boundary conditions $u(0) = 0$ and $u(r^*) = \frac{1}{2}$ at the free boundary. The conjugate flow B is found by a small perturbation of the primary flow A for the stream function. It can be written $U_B = U_A + v$. It turns out that v satisfies a linear second order differential equation whose solution can be computed explicitly in terms of Bessel functions. Here we always require the physical well-posedness, i.e. that the various physically relevant quantities circulation, total head, pressure and axial velocity are continuous across the free boundary.

This example is special because the bounded domain for the flow (considered initially) to be a sphere of radius R can be expanded to infinity. Here we have a justification of the approximation by bounded domains considered in our discussions of vortex rings of the last few sections. This approximation idea is very useful for a variety of problems of nonlinear science. We defer the details to another occasion.

Here there is a primary flow A that is supercritical for the parameter ω less than a critical value ω_c and subcritical otherwise. There is, in addition, a conjugate flow B that is subcritical and so gives rise to a vortex breakdown when the parameter exceeds the critical value ω_c. In fact, the stream function associated to B renders the associated Lagrangian, the flow force S, a minimax. On the other hand the primary flow A yields a minimum for S for the parameter ω sufficiently small.

Here is a quantitative discussion. The equation analyzed by Benjamin is

$$u_{rr} - r^{-1}u_r = \begin{cases} 2\omega^2 r^2 - 4\omega^2 u \\ \\ 0 \end{cases}$$

It arises from the Squire-Long equation directly above by setting

$$H'(u) = 2\omega^2 \quad \text{and} \quad I'(u) = 4\omega^2 u$$

We consider the solutions of this equation satisfying the two point boundary conditions $u(0) = 0$, $u(\xi) = 1/2$. The primary flow A has $\xi = 1$. The conjugate flow B has the free boundary slightly displaced to $\xi = 1 + \mu$.

The solution of the equation

$$u_{rr} - r^{-1}u_r = 2\omega^2 r^2 - 4\omega^2 u$$

together with the boundary condition $u(0) = 0$ is

$$u = \frac{1}{2}r^2 - \frac{1}{2}(Cr/\omega)\, J_1(2\omega r)$$

Here C is an arbitrary constant and J_1 is the first Bessel function. The boundary condition $u(\xi) = 1/2$ determines ξ and the constant C from the given physical data as well as the exact expression for the conjugate flow B. A crucial fact about the conjugate flow B is that, after elaborate calculation with Bessel functions, one finds the following expression for the flow force difference of the flows at A and B

$$S(B) - S(A) = K [\mu^3 + O(\mu^5)]$$

where K is a positive constant. This expression, of course, demonstrates that the flow B is subcritical.

Example 2

Another example considered by Benjamin consists of choosing

$$I'(u) = 2k^2u^2 \quad \text{and} \quad H'(u) = k^2u$$

The associated Squire-Long equation in one dimension with the z-variable omitted and setting $2y = r^2$ reduces to

$$d^2u/dy^2 = k^2 (u - u^2/y)$$

Here we use the two point boundary conditions $u(0) = 0$, $u(a) = a$. Thus in this example, the equation itself is truly nonlinear, so that explicit solutions are not to be expected and qualitative ideas are very important.

Here the parameter k will be determined large enough so that bifurcation occurs. A simple primary solution A is $u(y) = y$. To find the secondary solution bifurcating from A, we set $u = y + \phi$ so that ϕ satisfies

$$0 = \phi_{yy} + k^2(\phi + \phi^2/y)$$

together with zero boundary conditions at 0 and a. Standard bifurcation theory, as discussed in Chapter 2, yields a nontrivial solution for this equation with the zero boundary conditions occurring at the bifurcation point $\phi = 0$ and $k = \pi/a$. A computation of the relevant flow force S yields the fact that the primary flow A is supercritical for small values of $k < \pi/a$ and subcritical above the bifurcation, i.e., for $k > \pi/a$. The conjugate flow B cannot easily be exactly computed, but it is theoretically easy, using elementary bifurcation theory as described in Chapter 2.6, to say that the solution of the relevant boundary value problem exists for $k > \pi/a$. Moreover, it is fairly easy to see that although the conjugate states for $k > \pi/a$ are not unique, there is a conjugate state "adjacent" to the primary state $u = y$ that is subcritical for $k > \pi/a$.

These multiple states for a fixed Squire number (here denoted k) can be analyzed by the calculus of variations as follows: the relevant Lagrangian is called flow force, supercritical solutions of the reduced Squire-Long equation are distinguished by the fact that they represent minima of the flow force functional, and subcritical solutions are minimax critical points of this functional S. These solutions for a given physical problem are called a conjugate pair and the transition between the two states requires that the states be adjacent. This word "adjacent" when translated in mathematical language can be analyzed in terms of a norm for a solution, the difference of the two solutions being small but nonzero.

We now turn to a brief consideration of the more general vortex breakdown problem. This problem, although important, is still unresolved in its analytic details. A qualitative view is as follows: The major effects of swirling flow in vortex phenomena is that for large Squire numbers, bifurcation occurs in the associated free boundary problems. We illustrated this idea above for the second one-dimensional example discussed by Benjamin. The major fact is that a primary state for sufficiently large parameter value splits, that is, bifurcates into a primary motion and a secondary motion, the primary motion becoming subcritical after the bifurcation. The speculative point here is that new solutions are distinguished by the fact that the vorticity is concentrated in a bounded domain. Moreover, as the magnitude of the Squire parameter increases, we speculate that these confined vortices exhibit the desired supercritical-subcritical instability.

Other Types of Vortex Breakdown

As shown in the accompanying Figure, bubble breakdown considerations need to be supplemented to include the spiral type of breakdown. Actually there are numerous types of vortex breakdown that can occur in a fluid context. To analyze these considerations an analytical approach using alternate forms of symmetry supplementing axial symmetry proves most worthwhile.

In particular, instead of axial coordinates (r, z) it is possible to use helical coordinates (r, Λ) where the z coordinate is replaced by a single coordinate measured in a helical direction. In this case the Squire-Long equation needs to be modified and, in fact, we shall find in the sequel that it can be replaced by a helical generalization. This new modified Squire-Long equation then plays the same crucial role for helical coordinates that the standard Squire-Long equation plays for axisymmetric coordinates. The same methods of study for this nonlinear elliptic equation, namely, variational principle and Sobolev space, extend to this new context.

Helical coordinates depend on a parameter, namely, the winding of the helix around the axis of symmetry. Each of these new parameters gives rise to a modified type of vortex breakdown.

Mathematical Structures for the Squire-Long Equation

In this section, we wish to discuss the mathematical features of the Squire-Long equations that give rise to the rich vortex structure that appears in observation.

1) Elliptic Nature of the Squire-Long Equation

The Squire-Long equation in the axisymmetric coordinate system has as leading term the axisymmetric Laplace operator coupled with lower ordered terms that are nonlinear in nature but do not involve derivatives of the stream function. This sort of equation is a standard for the abstract mathematical theories of nonlinear analysis discussed in this book. The effect of this type of equation is that classical smooth solutions and weak solutions in a Sobolev space coincide. Thus, we use Sobolev techniques to discuss the Squire-Long equations in an appropriate Hilbert space context. See the solution diamond diagram of Appendix 3 of Chapter 1.

2) Calculus of Variations Formulation

The Squire-Long equation is special because the vortex solutions it contains arise from a variational principle that must be discovered ad hoc. This variational principle can be found by formal Hilbert space reasoning. This calculus of variations formulation enables us to formulate vortex solutions as minimax critical points of an associated Lagrangian energy functional.

3) Free Boundary Problems

The vortex solutions of the hydrodynamic equations arise as free boundary problems. Indeed, the notion of a vortex solution is based on the idea that the solenoidal part of the velocity vector field v, namely curl v, is confined to a <u>bounded</u> domain D in the fluid. The boundary of this domain ∂D is one of the key issues in the vortex problem, as it separates the solenoidal and irrotational parts of the fluid, i.e. it defines the region of vorticity in the fluid.

4) Bifurcation Phenomena and Parameter Dependence

The vortex motions determined by the Squire-Long equation depend crucially on a single parameter labelled S and known as the Squire parameter. When the parameter is less than a certain critical number, a certain type of vortex behavior is observed and when S is above a certain number, the observed vortex behavior is radically changed. In particular, Benjamin observed that in certain model cases, a certain supercritical vortex flow is changed into a subcritical vortex flow. An interesting distinction between these two vortex flows is as follows: supercritical steady flow cannot support infinitesimal standing waves while subcritical steady flow can support such waves. The distinguishing characteristic in the case of one spatial dimension is known as a "flow force" in Benjamin's terminology. The flow force in the case of one spatial dimension is simply the Lagrangian of the Squire-Long equation and is thus determined by the stream function as dependent variable.

New Classes of Solutions for the Squire-Long Equation

In studying vortex breakdown, we make use of new mathematical findings concerning vortex solutions of the Squire-Long equations. These results pertain to free boundary problems associated with the vorticity associated with the stream function in the case of axisymmetric Euler equations in three space dimensions. The results pertain to smooth steady solutions of the Euler equations that possess axial symmetry. The solutions in question vanish outside of a bounded smooth domain Ω in $\mathfrak{R}^3$. Inside of Ω the solution is strictly positive. The solution defines the vanishing or nonvanishing of a stream function for the fluid in question. The boundary of Ω denoted $\partial\Omega$ denoted the set separating the rotational part of the flow from the irrotational part. It is called the vortex core in question.

The new solutions discussed above, determined by new mathematical methods, show that the vortex rings discussed in Section 4.3 can be extended to vortex flows with swirl in a very direct manner. The calculus of variations questions involved yield a variational principle that is a characterization of the separated vortex flows involved. The solutions in question are not absolute minimizers, but rather minimize a kinetic energy subject to constraints. The associated Lagrange multiplier is the Squire parameter discussed above. Thus, the Squire parameter is determined a posteriori. By varying the kinetic energy via the parameter R, we find a family of vortex flows associated with the physical problem. By considering differences of solutions, we demonstrate bifurcation for certain critical values of the Squire parameter.

Vortex Cores and Vortex Breakdown

The first example discussed above is very important because it shows how vortex breakdown can occur in separated flow as a transition between a supercritical primary state and a secondary subcritical separated flow. In each of these flows the vorticity core is confined to a bounded domain outside of which the flow field is irrotational. The set of vorticity for each flow is called the vortex core. The mechanism for vortex breakdown is the transition from one type of vortex core to another type with the same flow characteristics, i.e. both flows are solutions of the same Squire-Long equations for an appropriately chosen Squire parameter. Yet the way in which the transition is made ,need not be assumed for the validity of the theory. Physically speaking the primary flow is supercritical, i.e. not permitting upstream traveling waves but permitting downstream wave propagation and at breakdown, criticality occurs so the flow allows standing waves. At subcriticality both upstream and downstream wave propagation can occur. The subcriticality has been the sole theoretical notion that correlates experimental data on vortex breakdown according to Leibovich. The mathematical mechanism for this transition is bifurcation. One goal of future analytic work is to show that the one-dimensional example described above can be extended to a two-dimensional case in which the axial coordinate is included.

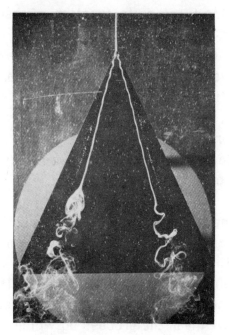

Section 4.7 Nonlinear Desingularization and Vortex Filaments

In Helmholtz's original paper on vorticity the notion of an idealized closed vortex filament of infinitely thin cross-section (or infinitely thin confinement domain for the Euler equations) is considered. Indeed, this idealization renders the confinement problem linear. Thus the stream function ψ for the (vortex filament) confinement problem, then satisfies a linear limiting case of the nonlinear equation (5) of Section 4.3. This linear limiting equation can be written

$$(1) \qquad \frac{\partial^2 \psi}{\partial r^2} - \frac{1}{r} \frac{\partial \psi}{\partial r} + \frac{\partial^2 \psi}{\partial z^2} = \delta(r - r_1)\delta(z - z_1)$$

where δ denotes the Dirac delta function. Consequently, ψ is a Green's function for the linear operator Δ on the right-hand side of the above equation, and so ψ has a singularity at (r_1, z_1). This last equation (as we shall see) arises as a linear limiting case of the nonlinear confinement problem described in the last few sections. However, this approximation leads to notable infinities and divergences for the speed of propagation of the associated circular vortex filament, so new ideas are needed to render the problem physically meaningful.

Now the degeneration of a "nonlinear" (smooth) vortex ring into a linear vortex filament representing a solution with an isolated singularity of a linear elliptic equation is a simple nontrivial example of the phenomenon of nonlinear desingularization we have mentioned many times in this book. More precisely, by "nonlinear desingularization" we mean the process by which a system of linear partial differential equations whose solutions possess singularities can be regarded as a degenerate form of a corresponding system of nonlinear equations whose solutions are smooth and moreover, possess smooth solutions that converge to the singular solutions of the linear system upon degeneration.

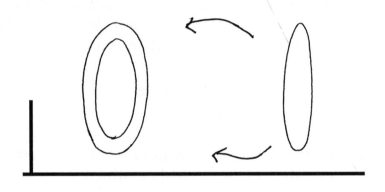

vortex ring   vortex filament

Nonlinear desingularization of vortex filament and vortex ring of small cross section.

To describe this situation, we consider

Problem 1   Give a coherent mathematically correct description of the degeneration of vortex rings into a circular vortex filament.

A number of mathematicians have attempted to resolve this problem by constructing asymptotic expansions about the Green's function (i.e., about the vortex filament) that represent thickened vortex rings. This approach is lacking, however, in treating the nonlinear problem on its own merits when no first approximation to the solution is available.

For a different, more fundamental approach, we consider the discussion in Section 4.4, and in particular the characterization of vortex rings given in the theorem of that section. Consider the isoperimetric variational problem conjugate to the problem (I) stated there for bounded domains $\pi(a, b)$, namely,

$$(W_R) \qquad\qquad \min_{\Sigma_R} \int_{\pi(a,b)} |\nabla\psi|^2 r\,dr\,dz$$

where Σ_R is the class of functions in the Sobolev space $W_{1,2}(\pi(a,b))$ that satisfies the integral constraint

$$\int_{\pi(a,b)} r^2 \tilde{F}(\psi - q)r\,dr\,dz = R$$

It is easy to show that this problem is equivalent to the variational problem (I) (i.e., a solution of the one problem leads to a solution of the other). Thus, it is natural to consider the behavior of the solutions ψ_R of the confinement problem associated with W_R as a function of R. In particular, to describe the behavior as ψ_R as $R \to 0$. In fact, one can prove

Theorem 2   Under the assumption of Theorem 1, let ψ_R denote any solution of the variational problem W_R with associated Lagrange multiplier λ_R, and with $q_R = 1/2\,Wr^2 + k_R$ and $k_R \to \infty$ set

$$h(R) = \lambda_R \int_{\pi(a,b)} r^2 f(\psi_R - q_R)r\,dr\,dz$$

Then as $R \to 0$, (i) the confinement domain $A_R = \{ (r,z) \, | \psi_R > q_R \}$ has diameter $\to 0$, and (ii) moreover, if $G(x, \overline{x}_R)$ denotes the Green's function relative to $\pi(a,b)$ and Δ for any $\overline{x}_R \in A_R$ and any bounded subdomain Ω of $\pi(a, b)$

$$\left\| \frac{\psi_R}{h(R)} - G(x, \overline{x}_R) \right\|_{L_R(\Omega)} \to 0$$

Corollary   Thus the family of vortex rings degenerates into a vortex filament, and the smooth stream function ψ_R converges to a solution of (L) that is proportional to the Green's function.

Some Key Ideas in the Proof

One of the main facts is to show that the confinement domain $A_R = \{ x \, | \psi_R > q \}$ converges to a point as $R \to 0$. (In fact, the need to establish this fact motivates our switch to the conjugate problem W_R). Here the appropriate domain functional to measure the size of A_R turns out to be a type of capacity defined as follows.

Let Ω be a domain in $\mathfrak{R}^N$, and A a domain strictly contained in Ω. Let a class of function $m(A)$ be defined as follows:

$$m(A) = \{ u \, | \, u \in W_{1,2} (\Omega), \quad u \geq 1 \text{ on } A \}.$$

Then the capacity of A relative to Ω, cap (A, Ω) is defined as

$$cap(A, \Omega) = \inf_{u \in m(A)} \int_{\Omega} |\nabla u|^2$$

Let A_R be defined as above, relative to the variational problem W_R discussed there. Then we can prove

Lemma $cap(A_R, \Omega) \to 0$ as $R \to 0$ and moreover $diam\ A_R \to 0$ as $R \to 0$.

This fact is proved by noting that on A_R, ψ_R the solution of the isoperimetric variational problem, satisfies a uniform lower bound from below, i.e. there is an absolute constant $\alpha > 0$ such that on A_R

$$\Psi_R \geq q_R(r,z) \geq \alpha > 0 ,$$

so $\psi_R/\alpha \geq 1$ on A_R. Consequently, $cap(A_R,\Omega) \to 0$ provided we show that the infimum in the variational problem $W_R \to 0$ with R. But this last fact is rather easy to prove. Moreover, since Ω is a bounded domain in $\mathcal{R}^2$ it is rather well known that since A_R is a connected set

$$cap(A_R, \Omega) \to 0 \quad as \quad R \to 0,$$

then $diam\ A_R \to 0$ follows by elementary analysis.

We now show that the fact that $diam\ A_R \to 0$ implies the Theorem , at least relative to convergence in $L_1(\Omega)$. Indeed, ψ_R satisfies

$$\Delta\Psi_R + \lambda_R \tilde{f}(\Psi_R - q_R) = 0$$

$$\Psi_R |_{\partial\Omega_{ab}} = 0 .$$

Consequently,

$$\Psi_R(x) = \lambda_R \int_{\partial\Omega_{ab}} G(x,z)\tilde{f}(\Psi - q_R)dV$$

$$= \lambda_R \int_{A_R} G(x,z)\tilde{f}(\Psi - q_R)dV \ ,$$

as so with a constant

$$h(R) = \lambda_R \int_{A_R} G(x,z)f(\Psi - q)dV$$

we find for $\bar{x} \in A_R$. Now as $R \to 0$ since ,

$$\text{diam } A_R \to 0, \quad \text{and} \quad \int_\Omega | G(x,\omega) - G(x,x_a)|dx \to 0$$

we find for $x_a \in A_R$,

$$\frac{\Psi_R}{h_R} - G(z,\omega) = \frac{\lambda}{h_R}\int_{A_R}\left\{ G(z,\omega) - G(z,x_a) \right\}f(\psi - q)$$

by virtue of elementary facts about Green's functions. Then, since
$h_R = \lambda \int_{A_R} \tilde{f}(\Psi_R - q_R)dz$, we find $\Psi_R/h_R \to G(z,\omega)$ in the norm
$L_1(\Omega)$, as required; actually, a much sharper estimate shows the convergence can be improved to $L_q(\Omega)$ convergence with any finite q.

Chapter 5 Mathematical Aspects of Superconductivity

Section 5.1 The Simplest Nonlinear Yang-Mills Theory that Works

Yang-Mills theory represents an attempt to understand subtle quantum phenomena via certain nonlinear partial differential equations arising as Euler-Lagrange equations of certain real-valued functionals that are invariant under a given gauge group G. Thus it seems natural to inquire whether there are known physical problems that can be successfully and quantitatively treated by such an approach and to explore (as fully as possible) the associated mathematical structures involved. In the linear case, the electrodynamics predicted by Maxwell's equations goes far beyond the goals just stated. Indeed, these linear partial differential equations served to form an unexpected unification in the dynamics of classical electricity, magnetism and optics. Correct predictions based on these linear equations have become a landmark in scientific history. Thus for nonlinear science, the topic discussed in this chapter is of the utmost importance.

Perhaps the simplest and most successful nonlinear theory of this type to date is that of V. Ginzberg and L. Landau, first proposed in 1950 to describe certain nonlinear macroscopic effects in superconductivity. The equilibrium states of this theory are determined by solving parameter dependent boundary value problems for a system of nonlinear elliptic equations combining a nonlinear Schrödinger equation and nonlinear version of Helmholtz's equation. The purely nonlinear mathematical aspects of the theory however yielded unexpected physical predictions for sufficiently large parameter values.

348

The Ginzberg-Landau equations were first derived as the Euler-Lagrange equations of the real-valued functional I_λ (called the Ginzberg-Landau Action Functional) defined by

$$I_\lambda = \frac{1}{2} \int_{\mathbb{R}^2} |dA|^2 + |D_A\phi|^2 + \frac{\lambda}{4}(1 - |\phi|^2)^2$$

where A is a one form defined on $\mathbb{R}^2$ with $A = (A_1, A_2)$ and ϕ, a complex valued scalar with real and imaginary parts denoted by subscripts $\phi = \phi_1 + i\phi_2$, $dA = \text{curl } A$, $D_A\phi = (d - iA)\phi$ the covariant derivative of ϕ with respect to A so that

$$|dA|^2 = (\partial_1 A_2 - \partial_2 A_1)^2 ,$$

$$|D_A\phi|^2 = |(d - iA)\phi|^2 =$$

$$(\partial_1\phi_1 + A_1\phi_2)^2 + (\partial_1\phi_2 - A_1\phi_1)^2 + (\partial_2\phi_1 + A_2\phi_2)^2 + (\partial_2\phi_2 - A_2\phi_1)^2 .$$

This functional was used in the original articles of Ginzberg and Landau in 1950 and forms the basis for the mathematical proofs we describe here. The functional I_λ is of significance because it is gauge invariant. This means that under the gauge transformation

$$(\phi, A) \rightarrow (\phi \exp(i\psi), A + \nabla\psi)$$

the functional is invariant. Thus the Ginzberg-Landau Action Functional represents a simple gauge theory with an Abelian gauge group $U(1)$, and it becomes an important problem to determine the appropriate equilibrium states (called here vortices in certain cases), that is, smooth critical points with boundary conditions (determined by the finiteness of I_λ) for this functional. It turns out that these critical points cannot be determined in closed form for arbitrary positive values of the parameter λ (here regarded as a real number). However, when $\lambda = 1$, this functional possesses an additional symmetry called "self-duality" and the second order partial differential equations of Ginzberg-Landau reduce to a first order system (see the discussion below).

In particular, Abrikosov, in a paper published in 1957, found that when the relevant parameter, λ say, exceeds this certain critical number $\lambda = 1$, new stable secondary solutions (termed "vortices") appear. Mathematically, these solutions turned out to be classified by an integer n related to $\pi_1(S')$ implying that discrete quantum states could be classified by homotopy theory, a situation that also occurs in higher dimensions. In contemporary science, high values of the parameter λ have proved extremely important (Type II superconductivity), and Abrikosov's pioneering predictions based on an interesting combination of physical and mathematical reasoning have proved of fundamental importance. For this critical parameter value $\lambda \equiv 1$, self-duality prevails and special methods may be used to find vortices.

Thus, when $\lambda > 1$ (the physically relevant case of Type II superconductivity), this self-duality property of the Ginzberg-Landau equation breaks down and new results are called for. These results are described below. One hint is Abrikosov's observation (based on physical reasoning) that in the limit as $\lambda \to \infty$, the radially symmetric vortex solution is given by the Green's function of an associated linear Helmholtz equation. In fact, we indicate below that for all nonzero integers n with $0 < \lambda < \infty$, there is a one parameter family of smooth radially symmetric solutions of the Ginzberg-Landau equations interpolating between the self-dual vortex solutions with $\lambda = 1$ and the $\lambda = \infty$ vortex solutions given by the Green's function mentioned above. Vortex solutions with $n > 1$ can be shown to be unstable in the sense that the second variation of the associated energy functions F_λ can be made negative definite, so that for $\lambda > 1$ the vortex solution with $n = 1$ is more stable, and in fact these $n = 1$ vortices are exactly the ones physically observed.

Mathematical Formulation and Statement of Results

The equilibrium states of the Ginzberg-Landau equations are the finite energy smooth critical points of the following action functional defined over $\mathcal{R}^2$.

$$(1) \qquad F_\lambda(\phi, A) = \int_{\mathcal{R}^2} \{ |\text{curl } A|^2 + \sum_j |(\partial_j - iA_j)\phi|^2 + \frac{\lambda}{4}(|\phi|^2 - 1)^2 \}$$

F_λ is called the Ginzberg-Landau action functional. The associated Ginzberg-Landau equations can be written as the nonlinear eigenvalue problem

$$\sum_{k=1}^{2} (\partial_k - iA_k)^2 \phi = \frac{\lambda}{2}(|\phi|^2 - 1)\phi$$

(1a)

$$(1b) \qquad \partial_k(\partial_k A_j - \partial_j A_k) = \text{Im}[\phi(\partial_j + iA_j)\overline{\phi}] \qquad k, j = 1, 2, \quad k \neq j$$

where ∂_k denotes $\partial/\partial x_k$, $i^2 = -1$ and λ is a positive parameter which depends on the superconducting material.

Here ϕ, known as the order parameter, is a complex scalar field and A is the vector potential of the magnetic field h so that $h = \text{curl } A$. The associated critical points can be classified by the "vortex number" n, an integer-valued topological invariant given by the integral formula

(2) $$n = \frac{1}{2\pi} \int_{\mathscr{R}^2} [\text{curl } A] dv$$

Finite energy smooth critical points of (1) (A,ϕ) with $n \neq 0$ are called "vortex solutions" (or vortices) in analogy with the theory of ideal fluids,since in this case, curl A, is an analogue of vorticity Indeed, if $n \neq 0$, h = curl A cannot vanish everywhere, is gauge invariant and is an analogue of vorticity. As mentioned above, these solutions appear to have the most physical relevance since when n = 1 they minimize the associated Gibbs free energy functional.

This finite-energy condition is attained very easily from the functional (1) by merely requiring that each term in the variational integral be in $L_2(R^2)$. In particular, this imposes restrictions on the behavior of the integrands of each of the three terms at infinity. Thus, as $|x|$ tends to infinity, one requires

(i) $|\psi|^2 \to 1$ (ii) $|\nabla_A \psi| \to 0$ (iii) $|F| = |dA| \to 0$

These conditions imply the following: first that

$$\psi \to e^{i\eta(\theta)} \quad \text{with} \quad \eta(\theta + 2\pi) = \eta(\theta) + 2\pi N$$

where N will be a (topologically invariant) integer, from (ii)

$$A \to -i\partial \ln \psi$$

These last two restrictions imply, loosely speaking, from the principle of the argument that by taking a line integral around a large contour $C_R = \{ |z| = R \}$ where R is very large that

$$N = (2\pi i)^{-1} \int_{CR} \psi'/\psi = (2\pi i)^{-1} \int_{CR} d\ln\psi$$

thus $R \to \infty$ via the above equations,

$$N = (2\pi)^{-1} \int_{R^2} \text{curl } A \qquad \text{(by Green's theorem)}$$

Continuous variations of ψ and A subject to the L_2 restrictions stated above do not change N. Thus N is (when its sign is appropriately adjusted) a topological invariant and can in fact be related to the homotopy group $\pi_1(S^1)$.

This topological invariant was first termed "flux quantization" by Abrikosov in the pioneering research article on Type II Superconductivity, mentioned above . In fact, the functional (1) had been used by the great Russian physicists Ginzberg and Landau in developing the theory of superconductivity from the point of view of nonlinear partial differential equations. However, it was Abrikosov who first discovered that when λ increased beyond 1 (a normalized reference number) that new physical phenomena occur due to topological invariant N. We will return to this aspect below.

In 1976 Bogomolnyi observed that the Ginzberg-Landau functional (1) for fixed $\lambda = 1$ exhibits self-duality symmetry mentioned above. To achieve this he rewrote (1) for arbitrary λ as follows

$$I_\lambda = \int_{R^2} [J_1^2(\psi, A) + J_2^2(\psi, A) + J_3^2(\psi, A)] + \int_{R^2} curlA + \frac{(\lambda-1)}{4} \int_{R^2} (|\psi|^2 - 1)^2$$

in terms of the flux-quantization invariant N, the second integral in the above expression is fixed and exactly equal to $2\pi N$. Thus if $\lambda = 1$ the so-called self-dual case, the last two integrals in the above formula are either constant or vanish Thus to find vortices in the self-dual case, Bogomolyni's formula above shows that it suffices to fix the vortex number N and then solve the first order system of nonlinear partial differential equations

$$J_1 = J_2 = J_3 = 0$$

Notice this minimizes I_λ when $\lambda = 1$ and flux quantization applies, according to the above formula. Here the different operators J_1, J_2, J_3 can be written as follows for $N > 0$, with $\phi = \phi_1 + i\phi_2$

$$J_1 = (\partial_1\phi_1 + A_1\phi_2) - (\partial_2\phi_2 - A_2\phi_1)$$

$$J_2 = (\partial_2\phi_1 + A_2\phi_2) + (\partial_1\phi_2 - A_1\phi_1)$$

$$J_3 = curlA + \frac{1}{2}(\phi_1^2 + \phi_2^2 - 1)$$

In fact, quantitative results on these vortex solutions are generally obtained by assuming $\lambda = \infty$ and studying solutions (tending to zero at infinity) of the linear Helmholtz equation

(3) $\Delta h - h = -2\pi n\delta(x)$,

where $\delta(x)$ is the Dirac delta function.

We now state two results on this situation.

Theorem 1 (On Vortex Families) For each finite $\lambda > 0$ and each integer n, the functional $F_\lambda(\phi, A)$ has a smooth radially symmetric vortex solution $(A_\lambda, \phi_\lambda)$, that is an absolute minimum of (1) among rotationally symmetric smooth finite energy fields (ϕ, A) with fixed vortex number n. Moreover, as $\lambda \to 1$, these vortex solutions coincide with radially symmetric self-dual solutions. (Here rotationally symmetric means $\underline{A} = S(r)d\theta$ and $\phi = R(r)e^{in\theta}$ with the boundary conditions $\begin{cases} R(0) = S(0) = 0 \\ |R(r)| \to 1 \text{ and } S(r) \to n \text{ as } r \to \infty. \end{cases}$

This ansatz due to Abrikosov will be discussed below.)

Theorem 2 (On Nonlinear Desingularization) As $\lambda \to \infty$ for each nonzero integer n, the radially symmetric vortex solutions (described in Theorem 1) yield $h_\lambda = \text{curl } A_\lambda$. This function, in turn, tends in the Sobolev space $W_{1,p}(\mathbf{R}^2)$ $(1 < p < 2)$ to the Green's function solution of (3) above.

In the sequel, we shall investigate these ideas as well as the proof of Theorems 1 and 2 more carefully, utilizing the ideas already discussed in this book.

Section 5.2   The Physical Viewpoint

Superconductivity concerns the behavior of special materials at low temperatures. The behavior of these materials is especially interesting under an external magnetic field. Three effects have become famous in the past fifty years.

Effect 1: The Meissner Effect Here an external magnetic field is exerted on a superconductive material. This material is characterized by a parameter λ. For such moderate values of λ the external magnetic field is excluded from the superconductivity.

Effect 2: The Appearance of Vortices For very large values of λ and of the external field H, and special superconducting materials, the magnetic field penetrates the superconductor in "flux" tubes, known as vortices. These materials are called "Type II Superconductors."

Effect 3: High Temperature Superconductivity For certain ceramic magnetic superconductive materials of Type II, it is known by ingenious experiments that such materials exhibit superconductivity at a much higher temperature than previously observed before 1987. This effect promises many new applications in modern life for such superconductivity.

Historical Review

Superconductivity was discovered by the Dutch physicist, K. Onnes, around 1911. This work was purely experimental in nature and required the use of very low temperature equipment. Onnes observed that the electrical resistance of such metals as lead, mercury and tin completely disappeared below a certain critical temperature that depended on the particular metal itself.

The first theoretical work was done by F. and H. London in the early 1930's and resulted in a simple extension of Maxwell's equations, using linear governing equations.

Initially, as mentioned earlier, linear theories were attempted to explain superconducting phenomena. But it soon became clear that a nonlinear theory would yield better explanations. In 1950, the Russian physicists Ginzberg and Landau extended this earlier work to a nonlinear theory involving a parameter λ and a scalar complex-valued function ϕ. The parameter λ measured the type of superconductive material involved.   Beginning a few years thereafter, Abrikosov discovered the appearance of vortices, theoretically from this Ginzberg-Landau model. His work involved little rigorous mathematics but did involve ingenious physical arguments and a lot of correct mathematical intuition.

A theory published in 1957 by Bardeen, Cooper and Schreiffer was an excellent development in this direction. The so-called BCS theory explained what was known at that time about superconductivity but also predicted new phenomena that were later confirmed by experiment. This microscopic theory is based on a coherent pairing of electrons resulting from a long range attraction.

A)   High Temperature Superconductivity

Superconductivity had been limited until 1987 in applicability because of the difficulty and cost of cooling the materials to very low temperatures, near absolute zero. A very exciting development occurred in 1987 that has added new interest to the field of superconductivity. Ceramic oxides were found to have definite indications of superconductivity at temperatures well above absolute zero. This has had the effect of extending the possibilities of superconductivity in real applications. In fact at the present time there is no coherent theory to explain these experimental developments, although vortex phenomena are thought to play a key role. It is known that superconductivity is a macroscopic quantum phenomenon. Thus, quantum effects are exhibited on a macroscopic scale rather than on a scale of atoms and molecules. It is this larger scale effect of superconductivity that adds particular interest to the subject.

Thus the new high temperature superconductors require a new theory. At present this theory is still being discussed. A few points are clear, however. The notion of vortices, (i.e. flux), and Type II superconductors that we shall mention later, seem to be crucial ideas in the new theory. The arguments of Abrikosov, Ginzberg and Landau, based as they are on a mean field type approach to superconductivity, are indeed interesting but need modification to survive in this new high temperature regime. We hope the discussion in this book may find readers who contribute to this great goal.

Section 5.3   The Linear Approach to Superconductivity and Nonlinear Desingularization

Before nonlinear structures were introduced into superconductivity, London described vortices with vortex number N via the model linear equation.

$$(1) \qquad \Delta \underline{H} - \underline{H} = -2\pi N \, \delta(x) \quad \text{in } \mathfrak{R}^2$$

$$H \to 0 \quad \text{at } \infty$$

Here $\delta(x)$ is the Dirac delta function and $\underline{H} = \text{curl } A$ is the magnetic field associated with electric potential A. Note that $\underline{H}$ is a gauge invariant vector function and, as such, is a physically observable quantity. This equation (1) is named after its discoverers, the London brothers, and can be derived directly from Maxwell's equation with the addition of one extra equation and an ad hoc assumption

(+) $H = -k \text{ curl } J$,   where k is a phenomenological constant

We couple this equation with one of Maxwell's equations

$$\text{curl } H = (4\pi/c)J$$

together with the gauge div H = 0 to find

$$\Delta H - c_L H = 0, \quad \text{where } c_L \text{ is a positive physical parameter}$$

To include vortices in the discussion, London was compelled ad hoc to add a delta function to the right-hand side of equation (+) above, in order to yield an equation similar to (1). This ad hoc inclusion of a delta function turns out to be correct physically. We now pursue the mathematical context of this idea.

Here we shall show that this equation (1), directly above, is the limit as $\lambda \to \infty$ of vortex solutions with fixed vortex number N of the Ginzberg-Landau equations for any finite λ. These equations, as we shall show, have vortex solutions for fixed N that are of finite action, finite-valued and smooth everywhere. Thus the process for vortices of letting the parameter λ pass from infinity, governed by the London equation (1), to a finite number (governed by the Ginzberg-Landau equations) we have called "nonlinear desingularization" in this book because the "singular" Green's function solution of (1) passes over to the smooth vortex (with integer N) solution of the Ginzberg–Landau equations.

To analyze the mathematical analysis behind these ideas we begin by choosing polar coordinates (r, θ) on $\mathfrak{R}^2$. We use Abrikosov's ansatz for symmetric vortices and set

(*) $\phi(r, \theta) = R(r)e^{iN\theta}$ for fixed N

$$A(r,\theta) = S(r)d\theta$$

with the boundary condition $R(0) = S(0) = 0$, and $R(r) \to 1$ $S(r) \to N$ as $r \to \infty$. Note that such vortices, if smooth, automatically have vortex number N.

Then the action functional can be rewritten as $I_\lambda(R,S)$ defined below (see equation (') of Section 5.4)

$$(') \quad I_\lambda(R, S) = \frac{1}{2}\int_{\mathfrak{R}^2} \left(\frac{1}{r}S'\right)^2 + (R')^2 + \frac{1}{r^2}(n - S)^2 R^2 + \frac{\lambda}{4}(1 - R^2)^2 \, .$$

The special form (*) is called an "ansatz" because it automatically restricts the vortices to have the given form. The ansatz (*) is a simplification of the problem of determining vortices with a fixed vortex number N. Indeed, if one minimizes $I_\lambda(A, \Phi)$ over finite energy admissible (A, Φ) one finds the desired absolute minimum of I_λ is zero attained by $\Phi \equiv 1$ and $A = 0$. If one simplifies I_λ by the ansatz (*) one finds the associated absolute minimum of $I_\lambda(R, S)$, defined by the equation above, is strictly positive and is attained, for fixed λ, by a smooth nontrivial vortex solution (R_λ, S_λ) of vortex number N that has the form (*). The mathematical proof of this fact is straightforward but long. It can be found in the author's paper (with Y.Y. Chen) mentioned in the bibliography and will be discussed in the next section. Here we analyze the key steps in the nonlinear desingularization argument. First, as in the case of vortices of ideal fluids, we derive a "vorticity" equation associated with the Ginzberg-Landau equations. We set H = curl A, and find an equation for H by taking the curl of A in the Ginzberg-Landau equations. The resulting equation is gauge invariant (since H is) and has the form

$$(**) \qquad\qquad -\Delta H + H = T_\lambda(\Phi, A, H)$$

where, after a computation, one finds

$$T_\lambda = H(1-(\phi_1^2 + \phi_2^2)) + A_1\partial_2(\phi_1^2 + \phi_2^2) - A_2\partial_1(\phi_1^2 + \phi_2^2) + 2\partial_1\phi_1\partial_2\phi_2 - 2\partial_2\phi_1\partial_1\phi_2$$

This equation (**) is derived by taking the curl of both sides in the original Ginzberg-Landau equation (1b) mentioned earlier in this chapter.

Note that T_λ does not depend explicitly on λ or the derivative of A, other than H itself. Now, an explicit computation, using the ansatz (*) gives the following facts.

Setting the "magnetic field" $H_\lambda = \text{curl } A_\lambda$ (i.e. $H_\lambda = \frac{1}{r} S_\lambda'(r)$ for $r \neq 0$), we will first prove that H_λ satisfies the reduced "vorticity equation"

(1)
$$-\Delta H_\lambda + H_\lambda = T_\lambda(R_\lambda, S_\lambda)$$
$$H \to \infty \quad \text{as} \quad |x| \to \infty$$

where we find from (*) that

(**) $$T_\lambda = \frac{2}{r} R_\lambda R_\lambda'(n - S_\lambda) + \frac{1}{r}(1 - R_\lambda^2)S_\lambda' \quad \text{for } r \neq 0$$

where, using this ansatz, we have used the facts

$$|dA|^2 = r^{-2}(S'(r))^2$$

and

$$|D_A \phi|^2 = (R'(r))^2 + r^{-2}(n - S(r))^2 R^2(r)$$

and moreover, we shall prove the following integral identity for the vortex solution (R, S) in the next section

(†)
$$\int_{\mathscr{R}^2} \left(\frac{1}{r} S'\right)^2 = \frac{\lambda}{4} \int_{\mathscr{R}^2} (1 - R^2)^2 .$$

In the next section, we shall use the direct method of the calculus of variations to prove that vortex solutions satisfying the Abrikosov ansatz (*) exist for each fixed integer N. Thus we shall assume, in the rest of this section, that the vorticity equation (1) has a solution H_λ. Before proceeding further, we derive properties for this solution.

First, we sketch the proof that $H_\lambda \to 0$ exponentially at ∞. To this end, it is sufficient to show that $H_\lambda \to 0$ at ∞ and $T_\lambda \to 0$ exponentially at ∞ because H_λ satisfies the elliptic equation (1). Indeed, $H_\lambda(r) = (1/r)S'_\lambda(r) \in L_2[1, \infty)$. Moreover the functions R(r) and S(r) satisfy the following nonlinear coupled system: Assuming (ϕ, A) is a symmetric solution of vortex number n of the Ginzberg–Landau equations on $\mathscr{R}^2 \backslash \{0\}$ if R and S satisfy

$$- R''(r) - \frac{1}{r} R'(r) + \frac{1}{r^2} (n - S(r))^2 R(r) + \frac{\lambda}{2} (R^2(r) - 1)R(r) = 0,$$

$$- S''(r) + \frac{1}{r} S'(r) - (n - S(r))R^2(r) = 0$$

Here r varies over the interval $(0, \infty)$, and in addition, we add the boundary conditions $R(0) = S(0) = 0$ and $R(r) \to 1$, $S(r) \to n$ as r tends to ∞. This system simply comes from the Abrikosov ansatz substituted into the Ginzberg–Landau action functional.

In the next sections we shall show that this system has smooth solutions satisfying the stated boundary conditions at zero and at infinity. Assuming this result, deducing from it the enhanced decay at infinity of R and S due to the nonlinearity of the system, and using the definition of T_λ in (**), we obtain the desired decay results.

Now one shows that

(i) $T_\lambda \geq 0$

(ii) $\displaystyle\int_{\mathcal{R}^2} T_\lambda dx = 2\pi N$ for all $\lambda > 0$

Here is the proof of (ii).   Assume vortex solutions with the Abrikosov ansatz (*) exist, as will be shown in the next section. Then, because $R^2(0) = 0$, $S(0) = 0$, $0 \leq R^2 < 1$, and $S \rightarrow n$ as $r \rightarrow \infty$, integration by parts gives, using (**) above yields

$$\int_{\mathcal{R}^2} \frac{2}{r} RR'(n - S)r \, dr \, d\theta = 2\pi \int_0^\infty (R^2)'(n - S)dr$$

$$= 2\pi \int_0^\infty R^2 S' dr,$$

$$= \int_{\mathcal{R}^2} \frac{1}{r} R^2 S',$$

Therefore, an amazing cancellation occurs if one uses (**)

$$\int_{\mathcal{R}^2} T = \int_{\mathcal{R}^2} \frac{1}{r} S' = 2\pi \int_0^\infty S' dr$$

$$= 2\pi(\lim_{k \to \infty} S(k) - S(0)) = 2\pi n.$$

The result (ii) is thus obtained.

To proceed further to consider very large values of the parameter λ, we use the integral identity (') mentioned above to investigate the behavior of the vortex solution as $\lambda \to \infty$. As we have shown above, the expression on the left-hand side of this identity is simply the square of the L_2 norm of the magnetic field H_λ. Provided we fix the vortex number n of the vortex solutions in question, as $\lambda \to \infty$, it is fairly easy to show that this quantity is uniformly bounded. From this fact, we conclude that as $\lambda \to \infty$,

$$\lim \int_{\mathcal{R}^2} (1 - R^2)^2 = 0$$

This means that R^2 tends to 1 in the L_2 sense. Actually, we can sharpen this fact to prove that

for all $\varepsilon > 0$, $R^2 \to 1$ uniformly on $[\varepsilon, \infty)$.

From this result we now show the following property of T_λ:

(iii)   For any $\varepsilon > 0$, $\lim_{\lambda \to \infty} \int_{|x| \geq \varepsilon} T_\lambda dx = 0$

Proof   Recall that for $r \neq 0$,

$$T_\lambda = \frac{2}{r} R_\lambda R_\lambda'(n - S_\lambda) + \frac{1}{r}(1 - R_\lambda^2)S_\lambda' .$$

By the facts already mentioned, it can be shown that

$$0 \leq \int_{|x| \geq \varepsilon} \frac{1}{|x|}(1 - R_\lambda^2)S_\lambda' \leq \int_{\mathfrak{R}^2} \frac{1}{|x|}(1 - R_\lambda^2)S_\lambda$$

$$= \int_{\mathfrak{R}^2}(1 - R_\lambda^2)H_\lambda \to 0 \quad \text{as } \lambda \to \infty .$$

On the other hand,

$$\lim_{\lambda \to \infty} \int_{|x| \geq \varepsilon} \frac{2}{|x|} R_\lambda R_\lambda'(n - S_\lambda) = 0 .$$

Indeed,

$$0 \leq \int_{|y| > \varepsilon} \frac{2}{|y|} R_\lambda R_\lambda'(n - S_\lambda)dy = 2\pi \int_\varepsilon^\infty (R_\lambda^2)'(n - S_\lambda)dr$$

$$\leq 2\pi n \int_\varepsilon^\infty (R_\lambda^2)' \quad \text{(by } 0 \leq S_\lambda < n \text{ and } (R_\lambda^2)' \geq 0)$$

$$= 2\pi n(1 - R_\lambda^2(\varepsilon)) \quad \text{(by } \lim_{r \to \infty} R_\lambda^2(r) = 1)$$

$$\to 0 \quad \text{as } \lambda \to \infty.$$

More generally,

(iv) $T_\lambda \to 2\pi N\delta(x)$ as $\alpha \to \infty$ in the integral sense.

We shall not prove this result here, but refer to the paper of Berger and Chen in the Bibliography.

We then prove that for fixed n, as $\lambda \to \infty$,

(2) $H_\lambda \to 2\pi nG$ in the Sobolev space $W_{1,p}(\mathbb{R}^2)$, $1 \leq p < 2$

where G is the Green's function satisfying the linear equation

(3) $\begin{cases} -\Delta G(x) + G(x) = \delta(x) \\ G(x) \to 0 \text{ as } |x| \to \infty \end{cases}$

where $\delta(x) \equiv$ Dirac delta function. Furthermore, it was shown in (ii) above that

(4) $\displaystyle\int_{\mathcal{R}^2} T_\lambda \equiv 2\pi n$ for all $\lambda > 0$

and thus it seems reasonable to expect, given (iii)

(5) $T_\lambda \to 2\pi n\delta(x)$ as $\lambda \to \infty$

in an integral sense. (Therefore, (3) is called a degenerate form of (1)).

We may assume the vortex number $n > 0$ from now on.

A few words are given here in order to derive an intuitive idea concerning the limiting process of H_λ. In fact, the Green's function G is proportional to the zero-order Hankel function of imaginary argument, i.e., the second kind of modified Bessel function of order zero $K_0(r)$. It is known that K_0 is positive and decreasing in (0,

∞), tends toward infinity at the origin and toward zero exponentially at infinity. On the other hand, from (*) above, we find

$$H_\lambda = \frac{1}{r} S_\lambda' \geq 0 \quad \text{and} \quad H_\lambda' = \frac{1}{r} S_\lambda'' - \frac{1}{r^2} S' = -\frac{1}{r}(n-S_\lambda)R_\lambda^2 < 0 \quad \text{in } (0, \infty)$$

and H_λ tends toward zero exponentially at infinity. (See Figure 1).

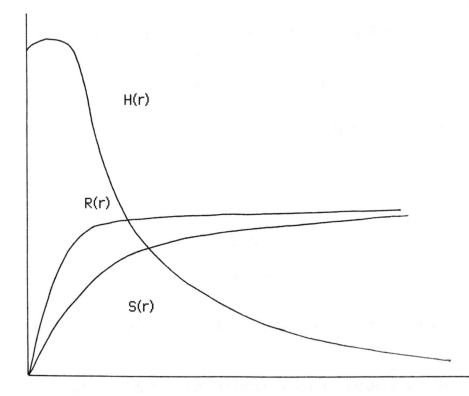

FIGURE 1   Graphs of R(r), S(r), and H(r) for vortex number n = 1, and fixed large λ. (Note $K_0(r)$ tends to infinity as r tends to zero, but for finite r resembles H(r) closely.)

Now, proceeding further with these ideas, the desingularization phenomenon can be expressed mathematically by the result

Theorem As $\lambda \to \infty$, the vortex solution for H_λ for fixed N, obtained above by minimizing $I_\lambda(R,S)$ with the ansatz (*) over appropriate Sobolev spaces for R and S satisfies

$$H_\lambda(x) \to 2\pi N\, G(x) \quad \text{in } W_{1,p}(\mathbb{R}^2) \text{ for } 1 \le p < 2$$

(Sketch of Proof) This result is established in the author's joint paper referred to above. For the purpose of the text here we describe the main points of the argument. First, by (1), we have the representation

$$H_\lambda(x) = \int_{\mathbb{R}^2} G(x-y)\, T_\lambda(y)\, dy$$

where G is the appropriate Green's function for the operator $(\Delta - 1)$ on $\mathbb{R}^2$. Consequently, one can estimate Sobolev norms of the left hand side of the above equation by estimating appropriate Sobolev norms of the difference $\{G(x-y) - G(x)\}$. For the Green's function $G(x)$ in question, such results are well known.

Hence, by the property (ii) of T_λ, $\int_{\mathbb{R}^2} T_\lambda = 2\pi n$

Thus,

$$H_\lambda(x) - 2\pi nG(x) = \int_{\mathcal{R}^2} (G(x - y) - G(x))T_\lambda(y)dy$$

To proceed further with the Sobolev estimates for the left-hand side of the above equation, we now note that we show here how to find Sobolev estimates for $(G(x - y) - G(x))$ in terms of y. Once these estimates are found, appropriate use of Hölder's inequality completes the proof desired.

To find Sobolev estimates for the Green's function of the Helmholtz equation, we note that $G(x - y) = (1/2\pi)K_0(|x - y|)$ where K_0 is the modified Bessel function of zero order of the second kind. This function has been carefully studied. In fact, it is known that

$$K_0(r) = -(\ln r - \ln 2 + \gamma) I_0(r) + P(r)$$

where γ is Euler's constant, $I_0(r)$ is the modified Bessel function of order zero of the first kind and $P(r)$ is a convergent power series in r. Using this fact, we can prove the following Sobolev estimate:

For given $1 \leq p < 2$, there exists a constant $C > 0$, such that for all $y \in \mathcal{R}^2$,

$\|G(x - y)\| \leq C$ where the norm is the Sobolev norm $W_{1,p}(\mathcal{R}^2)$.

These estimates represent the first key steps in obtaining a rigorous proof of the nonlinear desingularization result in the case of superconducting vortices. For full details we refer the reader to the paper of Berger and Chen in the Bibliography.

Section 5.4   Function Spaces for Symmetric Vortices

It remains to demonstrate that symmetric vortex solutions for the Ginzberg-Landau equations exist for arbitrary positive λ using Abrikosov's ansatz for fixed vortex number n. In this section we specify the Sobolev spaces that will be used in the variational characterization of these vortices. In the final section, we use the direct method of the calculus of variations to show the desired vortices actually exist.

By a symmetric vortex of the Ginzberg-Landau equations, using Abrikosov's ansatz, and fixing the vortex number n, we mean a smooth finite-action solution that can be written, in terms of polar coordinates (r, θ) on $\mathbb{R}^2 \backslash \{0\}$, as

$$\phi(r, \theta) = R(r)e^{in\theta}$$
$$A(r, \theta) = S(r)d\theta$$

together with the boundary conditions

$$R(0) = S(0) = 0$$

and

$$R^2(r) \to 1, \; S(r) \to n, \; \text{as} \; r \to \infty$$

where n is a nonzero integer, called the vortex number. We note that if $A = Sd\theta$, then $dA = S'drd\theta$ and $|dA|^2 = |\frac{1}{r}S'|^2$.

We begin by attempting to find closed subsets of Hilbert spaces for functions R and S, with fixed integer n, such that the pair (ϕ, A) determined by the relevant Euler-Lagrange equations renders $I_\lambda(\phi, A)$ finite. The reduced functional of $I_\lambda(\phi, A)$ obtained by utilizing the given ansatz is

$$(\dagger) \quad I_\lambda(R, S) = \frac{1}{2} \int_{\mathcal{R}^2} \left(\frac{1}{r} S'\right)^2 + (R')^2 + \frac{1}{r^2}(n - S)^2 R^2 + \frac{\lambda}{4}(1 - R^2)^2 .$$

We now define the function spaces for the real-valued radially symmetric function R and S. Let

C_R = the set of real-valued radially symmetric functions $R(|x|)$ defined on $\mathcal{R}^2$, such that R is nonnegative almost everywhere and $1 - R \in W_{1,2}(\mathcal{R}^2)$.

C_S = the set of real-valued radially symmetric functions $S(|x|)$ defined on $\mathcal{R}^2$, such that $\frac{1}{r} S \in L_2^{loc}(\mathcal{R}^2)$ with $\frac{1}{r} S' \in L_2(\mathcal{R}^2)$ where the derivative S' is in the distributional sense.

Note that elementary analysis shows the following

Fact C_S is a Hilbert space with the norm $\|S\| = \|(1/r)S'\|_{L_2(\mathcal{R}^2)}$.

Remark 1   The reason for choosing $R \geq 0$ is as follows. Let $\tilde{C}_R$ be the set of real-valued radially symmetric functions $R(|x|)$ defined on $\mathcal{R}^2$ which satisfies $R \in L_2^{loc}(\mathcal{R}^2)$, $R' \in L_2(\mathcal{R}^2)$ and $1 - |R| \in W_{1,2}(\mathcal{R}^2)$.   Because $C_R \subseteq \tilde{C}_R$,   we   have $\inf_{\tilde{C}_R} I_\lambda(R, S) \leq \inf_{C_R} I_\lambda(R, S)$. On the other hand, if $R \in \tilde{C}_R$, then $|R| \in C_R$ and

$$I_\lambda(|R|, S) \leq I_\lambda(R, S)$$

thus $\inf_{C_R} I_\lambda(R, S) \leq \inf_{\tilde{C}_R} I_\lambda(R, S)$. Moreover, $I_\lambda(R, S) = I_\lambda(-R, S)$ and we can show that if (R, S) is a minimizing solution of I_λ over $\tilde{C}_R$, then either R is nonnegative with $R(r) \to 1$ as $r \to \infty$ or R is nonpositive with $R(r) \to -1$ as $r \to \infty$.

Remark 2 Because in Cartesian coordinates $A_1 = \dfrac{-x_2 S}{x_1^2 + x_2^2}$ and

$A_2 = \dfrac{x_1 S}{x_1^2 + x_2^2}$, the smoothness of A at the origin forces $S(0) = 0$.

This is why we have the boundary condition $S(0) = 0$ and we set $\dfrac{1}{r} S \in L_2^{loc}(\mathbb{R}^2)$ in C_S.

Remark 3 We notice that the finite action elements of $C_R \oplus C_S$ are nonempty. That is, there is $(R, S) \in C_R \oplus C_S$ such that $I_\lambda(R, S) < \infty$. For example, let $R(|x|)$ be a C^∞ function on $\mathbb{R}^2$ such that $R(|x|) = 0$ for $|x| < 1$ and $R(|x|) = 1$ for $|x| > 2$, and define $S(|x|) = nR(|x|)$, then $R \in C_R$, $S \in C_S$ and $I_\lambda(R, S) < \infty$.

Lemma (On the vortex number) For $R \in C_R$, $S \in C_S$ and $\dfrac{1}{r} S' \in L_1(\mathbb{R}^2)$, the finite action of $I_\lambda(R, S)$ implies

$$\frac{1}{2\pi} \int_{\mathbb{R}^2} dA = \int_0^\infty S' dr = n$$

where n is the integer in the ansatz.

Proof   Since $\frac{1}{r}S' \in L_1(\mathcal{R}^2)$ for any sequence $\{k_j\}$ with $k_j \to \infty$,

$$\int_0^\infty S'dr = \frac{1}{2\pi}\int_{\mathcal{R}^2}\frac{1}{r}S' = \frac{1}{2\pi}\lim_{k_j\to\infty}\int_{B_{k_j}}\frac{1}{r}S' = \lim_{k_j\to\infty}\int_0^{k_j}S'dr$$

$$= \lim_{k_j\to\infty}[S(k_j) - S(0)] = \lim_{k_j\to\infty}S(k_j)$$

because of the Stokes theorem and $S(0) = 0$.   On the other hand, $I_\lambda(R, S) < \infty$ implies that there exists a sequence $\{k_j\}$ with $k_j \to \infty$, such that

$$0 = \lim_{k_j\to\infty}\int_{r=k_j}(1 - R^2)^2 r^2 d\theta = \lim_{k_j\to\infty}2\pi(1 - R^2(k_j))^2 k_j^2$$

as well as

$$0 = \lim_{k_j\to\infty}\int_{r=k_j}(n - S)^2 R^2 d\theta = \lim_{k_j\to\infty}2\pi(n - S(k_j))^2 R^2(k_j).$$

The first limit implies $\lim_{k_j\to\infty}R(k_j) = 1$. Combining this result with the second limit gives $\lim_{k_j\to\infty}S(k_j) = n$.

We now establish an integral identity for the minimizing solutions, which is used in the above Section 5.3.

Lemma   If (R, S) is an infimum of I_λ over $C_R \oplus C_S$, then

$$\int_{\mathscr{R}^2} \left(\frac{1}{r}S'\right)^2 = \frac{\lambda}{4}\int_{\mathscr{R}^2} (1-R^2)^2 \ .$$

Proof   Making the change of variable $c = \alpha r$ in $R(r)$ and $S(r)$, we define function $I(\alpha)$ from $(0,2)$ to $\mathscr{R}$ by

$$I(\alpha) = I_\lambda(R(\alpha r), S(\alpha r))$$

$$\frac{d}{d\alpha}I(\alpha) = \int_{\mathscr{R}^2} \frac{\alpha}{c^2}(S'(c))^2 - \frac{\lambda}{4\alpha^3}(1-R^2(c))^2 \quad \text{where } c = \alpha r.$$

Clearly, $R(\alpha r) \in C_R$ and $S(\alpha r) \in C_S$ for $\alpha \in (0,2)$. thus $I(\alpha) \geq I(1)$ for all $\alpha \in (0,2)$. This gives $\frac{d}{d\alpha}I(1) = 0$, that is,

$$\int_{\mathscr{R}^2} \left(\frac{1}{r}S'\right)^2 = \frac{\lambda}{4}\int_{\mathscr{R}^2} (1-R^2)^2 \ .$$

Section 5.5   The Existence of Critical Points for I_λ Associated with Symmetric Vortices

Now: (ϕ, A) is a symmetric solution of vortex number n of the Ginzberg–Landau equations on $\mathcal{R}^2 \backslash \{0\}$ if R and S satisfy

$$-R''(r) - \frac{1}{r} R'(r) + \frac{1}{r^2} (n - S(r))^2 R(r) + \frac{\lambda}{2} (R^2(r) - 1) R(r) = 0,$$

(1)

$$-S''(r) + \frac{1}{r} S'(r) - (n - S(r)) R^2(r) = 0$$

Here r varies over the interval $(0, \infty)$, and in addition, we add the boundary conditions $R(0) = S(0) = 0$ and $R(r) \to 1$, $S(r) \to n$ as r tends to ∞. This result is simply obtained by deriving the Euler–Lagrange equations for the functional $I(R,S)$ of the last section.

To find these symmetric solutions of vortex number n for every parameter value in the interval $(0, \infty)$ by virtue of the discussion in the previous sections, we use the calculus of variations. Indeed, the solution diamond diagram of the Appendix at the end of Chapter 1, shows that once we find a finite action critical point using the Abrikosov ansatz for each vortex number n, fairly standard regularity theory for the smoothness of the pair (R, S) yields a solution of the system (1) directly above.

Thus we find a critical point of $I_\lambda(R, S)$ in the space $C_R \oplus C_S$ given by the expression

$$(^1) \quad I_\lambda(R, S) = \frac{1}{2} \int_{\mathscr{R}^2} \left[\left(\frac{1}{r} S' \right)^2 + (R')^2 + \frac{1}{r^2} (n - S)^2 R^2 + \frac{\lambda}{4} (1 - R^2)^2 \right].$$

associated with symmetric (finite-action) vortices of fixed vortex number n. Indeed, the smooth critical points of this functional are precisely the solutions of the coupled system of differential equations (1). The proof of this regularity result can be found in the paper of Berger and Chen mentioned in the Bibliography.

First we note that the solution of the Ginzberg-Landau equations obtained by setting $\phi \equiv 1$ and $A \equiv 0$ is a trivial solution rendering $I_\lambda(\phi, A) = 0$ without restricting the vortex number n to be nonzero.

Secondly, we note that with the restrictions imposed on (R, S),

$$(*) \qquad \alpha = \inf_{C_R \otimes C_S} \; I_\lambda(R, S) > 0 \qquad \text{for n nonzero}$$

Proof: Otherwise $I_\lambda(R, S) = 0$ and, from the definition of $I_\lambda(R, S)$ given in $(^1)$

$$\int_{\mathscr{R}^2} (1 - R^2)^2 = 0 \quad \text{and} \quad \int_{\mathscr{R}^2} (1/r^2)(n - S)^2 R^2 = 0$$

These two equations imply that $R^2 = 1$ and that $S = n \neq 0$. But this contradicts the fact that $\frac{1}{r} S \in L_2^{loc}(\mathscr{R}^2)$.

The main result of this section is stated in the following result.

Theorem   The infimum, α, of I_λ over the Hilbert space described above is attained. Moreover, the infimum α is strictly positive.

The proof of the attainment of the infimum will be carried out in several standard lemmas.

First, a result on the coerciveness of $I_\lambda(R, S)$

Lemma 1 $I_\lambda(R_k, S_k) \to \infty$ whenever $\|1 - R_k\|_{W_{1,2}(\mathcal{R}^3)} + \|S_k\|_{C_s} \to \infty$,

where $R_k \in C_R$ and $S_k \in C_s$.

Proof of Lemma 1   This result is an immediate consequence of the following estimate

$$\| 1 - R \|^2_{W_{1,2}(\mathcal{R}^3)} + \|S\|^2_{C_s} = \|1 - R\|^2_{L_2} + \| R'\|^2_{L_2} + \|1/r\, S'\|^2_{L_2}$$

$$\leq \max\left\{\frac{8}{\lambda}, 2\right\} I_\lambda(R, S)$$

Finally, a result on the lower semicontinuity of $I_\lambda(R, S)$ with respect to weak convergence in $C_R \oplus C_s$.

Lemma 2   Let $\{R_k, S_k\}$ be a sequence in $C_R \oplus C_s$ and (R, S) and element in $C_R \oplus C_s$. Suppose that

(I)
$$1 - R_k \to 1 - R \text{ weakly in } W_{1,2}(\mathcal{R}^2)$$

and

(II)
$$S_k \to S \text{ weakly in } C_s.$$

Then

$$I_\lambda(R, S) < \lim_{k \to \infty} I_\lambda(R_k, S_k) .$$

Proof of Lemma 2   By the hypotheses of the lemma, we have

$$\int_{\mathcal{R}^2} (R')^2 + (1/r\, S')^2 \leq \lim_{k \to \infty} \int_{\mathcal{R}^2} (R'_k)^2 + (1/r\, S'_k)^2$$

From the definition of weak convergence, it is easy to see that $1 - R_k(r) \to 1 - R(r)$ weakly in $W_{1,2}(0, L)$ and $S_k(r) \to S(r)$ weakly in $W_{1,2}(0, L)$ for all $L \to \infty$. Hence, according to the Sobolev-Kondrachov embedding theorem $R_k(r) \to R(r)$ and $S_k(r) \to S(r)$ almost   everywhere.   The   result   now   follows   by Fatou's Lemma.

Proof of the Theorem   We may assume that $\{(R_k, S_k)\}$ is a minimizing sequence in $C_R \oplus C_S = H$.

$$\lim_{k \to \infty} I_\lambda(R_k, S_k) = \inf I_\lambda(R, S) \leq \alpha < \infty$$

From the coerciveness lemma, $\{1 - R_k\}$ and $\{S_k\}$ are bounded sets in $W_{1,2}(\mathcal{R}^2)$ and C_S respectively. We may assume $1 - R_k \to 1 - R$ weakly in $W_{1,2}(\mathcal{R}^2)$ where $R \in C_R$. Since C_S is a Hilbert space and $\{S_k\}$ is a bounded set in it, we may assume $S_k \to S$ weakly in C_S where $S \in C_S$. Using the lower semicontinuity lemma, we obtain

$$I_\lambda(R, S) \leq \lim_{k \to \infty} I_\lambda(R_k, S_k) = \inf_{C_R \oplus C_S} I_\lambda(R, S)$$

Since $(R, S) \in H$, we have $\quad I_\lambda(R, S) = \inf_H I_\lambda(R, S)$.

The displayed equation (*) in the beginning of this section gives $I_\lambda(R, S) > 0$. Thus the Theorem stated immediately above is proven.

Additional Facts

(i)   The minimizing solution (R, S) obtained above for each fixed λ and for each nonzero integer n is a generalized solution of the system of nonlinear ordinary differential equations (1) mentioned at the beginning of this section.

(ii)   The minimizing solutions (R, S) are smooth. In fact, the functions are elements of the space $C^2(0, \infty)$, and in fact the associated quantities (A, ϕ) can be extended to the whole plane in a C^2 manner. (For a proof see the above-mentioned paper of Berger and Chen.)

(iii)   We now mention the properties of the extremal smooth solution (R, S) for fixed vortex number $n \neq 0$. The main results are

(a)   sharp a priori bounds for R and S, namely for all $r \in (0, \infty)$

$$0 < R^2 < 1, \text{ and } 0 < S(r) < n.$$

(b) $S(r)$ and $R(r)$ are strictly increasing functions
(c)   asymptotic behavior of R and S as $r \to \infty$; namely
        $R(r) \to 1$ and $S(r) \to n$ exponentially as $r \to \infty$.

(iv) The nonlinear eigenvalue problem described above is quite unusual in this case by virtue of the Abrikosov ansatz. Indeed, this ansatz, by virtue of the results described above, reduces the search for vortices of the Ginzberg-Landau equations to a strict minimization problem on a Hilbert space. A remarkable simplification.

Bibliography

Chapter 1

Arnold, V. I., Mathematical Methods of Classical Mechanics, Springer-Verlag, 1978.

Arnold, V. I., Geometrical Methods in the Theory of Ordinary Differential Equations, Springer-Verlag, 1983.

Berger, M. S. and Berger, M. S., Nonlinearity and Functional Analysis, Academic Press, 1977.

Berger, M. S. and Church, P., "Complete integrability and perturbation of a nonlinear Dirichlet problem," Indiana Journal of Mathematics, pp. 935-952, Nov/Dec 1979.

Drazin and Johnson, Solitons, Cambridge University Press, 1989.

McKean, H. P. and Scovil, J. C., "Geometry of Some Simple Nonlinear Differential Operators," Annali Scuola Normale Superiore, Pisa, Serie iv, Vol. XIII, pp. 299-346, 1986.

Schechter, M., Functional Analysis, Academic Press, 1972.

Watson, G. N., Theory of Bessel Functions, Cambridge University Press, 1935.

Moser, J., Stable and Random Motions in Dynamical Systems, Annals of Mathematics Studies, Princeton University Press, 1973.

Chapter 2

Berger, M. S. and Berger, M. S., Perspectives in Nonlinearity, Benjamin Publishers, 1968.

Berger, M. S., Nonlinearity and Functional Analysis, Academic Press, 1977.

Berger, M. S. and Bombieri, E., "On Poincaré's Isoperimetric Variational Problem for Closed Simple Geodesics," Journal of Functional Analysis, Vol. 42, pp. 274-298, 1981.

Berger, M. S. and Fraenkel, L. E., "Nonlinear desingularization in certain free boundary problems," Communications of Mathematical Physics, vol. 77, 00. 149-172, 1980.

Bombieri, E., An Introduction to Minimal Currents and Parametric Variational Problems, Mathematical Reports, 1985.

Feigenbaum, M. J., "Universal metric properties of nonlinear transformations," J. Stat. Physics, vol. 21, p. 669 (1979).

Poincaré, H., On Geodesics on Ovaloids, Trans. Amer. Math. Soc., 1901.

Chapter 3

Aubin, T., Nonlinear Analysis on Manifolds, Springer-Verlag Publishers, 1982.

Berger, M. S., "Simple Closed Geodesics and the Calculus of Variations," Annals of Math. Studies on Minimal Submanifolds, pp. 257-265, 1983.

Berger, M. S. and Bombieri, E., "On Poincaré's Isoperimetric Variational Problem for Closed Simple Geodesics," Journal of Functional Analysis, Vol. 42, pp. 274-298, 1981.

Berger, M. S. and Cherrier P., "A New Variational Method for Finding Einstein Metrics on Compact Kähler Manifolds," Journal of Functional Analysis, Vol. 79, pp. 103-135, 1988.

Besse, A., Einstein Manifolds, Springer-Verlag, 1987.

Bombieri, E., An Introduction to Minimal Currents, Harwood Academic Publishers, 1985.

Yamabe, H., "On a deformation of Riemannian structures on compact manifolds," Osaka Math. J. vol.12, pp. 21-37, 1960.

Wente H.,"Counterexample to a conjecture of H.Hopf, Pacific J. of Math. , 121 pp.193-243 (1986)

Chapter 4

Batchelor, G. K., An Introduction to Fluid Dynamics, Cambridge University Press, 1967.

Benjamin, T. B., "A theory of vortex breakdown," J. Fluid Mech., pp. 593-621, 1962.

Berger, M. S., Remarks on Vortex Breakdown, Chapter 15 in volume Vortex Dynamics (editor J. Calfleisch) published by SIAM, 1989.

Berger, M. S. and Fraenkel, L. E., "On the global vortex rings in an ideal fluid," Acta Mathematica, Vol. 132, pp. 13-51, 1974.

Lamb, H., Hydrodynamics, Dover Publishers, 1945 (originally published 1879).

Moffatt, H. K., "The degree of knottedness of tangled vortex lines," Fluid Mech., vol 35, part I, pp. 117-129, 1969.

Squire, H. B., Imp. Coll. Rep. 102, 1960.

Chapter 5

Berger, M. S. and Chen, Y. Y., "Symmetric vortices for the nonlinear Ginzberg-Landau equations of superconductivity and the nonlinear desingularization phenomenon," J. of Functional Analysis, vol. 82, pp. 259-295, 1989.

Jaffe, A. and Taubes, C., Vortices and Monopoles, Birkhauser, 1980.

London, F., Superfluids, vol. 1, John Wiley & Sons, 1950.

Tinkham, M., Introduction to Superconductivity, McGraw-Hill, 1975.

Problems

A Collection of Diverse Problems – mostly elementary, based on material of the text

Problems on Iteration

(i) Prove the following mappings $x_{n+1} = f(x_n)$ can be solved exactly by the formula given.

(ii) Verify the fixed points x^* as stated and the periodic points in each case.

a.

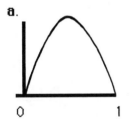

$$x_{n+1} = 4x_n(1 - x_n), \quad 0 \leq x_n \leq 1$$

$$x_n = \sin^2 (2^n \sin^{-1} \sqrt{x_0})$$

$$x^* = 0, \ 3/4$$

periodic if $x_0 = \sin^2\left(\pi\frac{q}{p}\right)$

($q < p$ integers).

b.

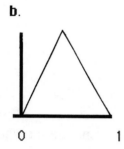

baker's transformation

$$x_{n+1} = \begin{cases} 2x_n & 0 \le x_n < \frac{1}{2} \\ 2(1-x_n) & \frac{1}{2} \le x_n \le 1 \end{cases}$$

$$x_n = \frac{1}{\pi} \cos^{-1}(\cos(2^n \pi x_0))$$

$$x^* = 0, \frac{2}{3}$$

periodic if $x_0 = \frac{q}{p}$

($q < p$ integers)

c.

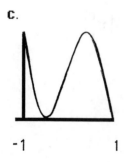

$$x_{n+1} = x_n(3 - 4x_n^2) \qquad -1 \le x_n \le 1$$

$$x_n = \sin(3^n \sin^{-1} x_0)$$

$$x^* = 0, \pm\frac{1}{\sqrt{2}}$$

periodic if $x_0 = \sin\left(\pi\frac{q}{p}\right)$

($q < p$ integers)

d. baker's transformation of the second kind

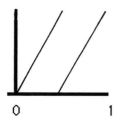

$$x_{n+1} = \begin{cases} 2x_n & 0 \le x_n \le \frac{1}{2} \\ 2x_{n-1} & \frac{1}{2} < x_n < 1 \end{cases}$$

$$x_n = \frac{1}{\pi}\cot^{-1}(\cot(2^n \pi x_0)) \quad 0 \le x_n \le 1$$

$$x^* = 0, 1$$

periodic if $x_0 = \frac{q}{p}$

($q < p$ integers)

e.

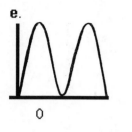

0   1

$$x_{n+1} = 16x_n(1 - x_n)(1 - 2x_n)^2 \quad 0 \leq x_n \leq 1$$

$$x_n = \sin^2(4^n \sin^{-1}\sqrt{x_0})$$

$$x^* = \frac{3}{4} \ , \quad \frac{(5 \pm \sqrt{5})}{8}$$

periodic if $x_0 = \sin^2\left(\pi\frac{q}{p}\right)$

(q < p   integers)

f.

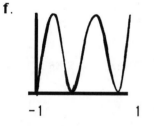

-1   1

$$x_{n+1} = x_n(5 - 20x_n^2 + 16x_n^4) \quad -1 \leq x_n \leq 1$$

$$x_n = \sin(5^n \sin^{-1}x_0)$$

$$x^* = 0 \ , \quad \pm\frac{1}{2}, \quad \pm 1$$

periodic if $x_0 = \sin\left(\pi\frac{q}{p}\right)$

(q < p   integers)

g.

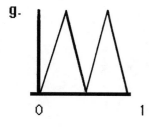

$$x_{n+1} = \begin{cases} 4x_n & 0 \le x_n < \dfrac{1}{4} \\[2mm] 2(1-2x_n) & \dfrac{1}{4} \le x_n < \dfrac{1}{2} \\[2mm] 2(2x_n-1) & \dfrac{1}{2} \le x_n < \dfrac{3}{4} \\[2mm] 4(1-x_n) & \dfrac{3}{4} \le x_n \le 1 \end{cases}$$

$$x_n = \frac{1}{\pi} \cos^{-1}(\cos(4^n \pi x_0))$$

$$x^* = 0, \frac{2}{5}, \frac{2}{3}, \frac{4}{5}$$

periodic if $x_0 = \dfrac{q}{p}$

($q < p$ integers)

h.

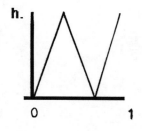

$$
x_{n+1} = \begin{cases} 3x_n & 0 \le x_n < \dfrac{1}{3} \\[2mm] 2-3x_n & \dfrac{1}{3} \le x_n < \dfrac{2}{3} \\[2mm] -2+3x_n & \dfrac{2}{3} \le x_n \le 1 \end{cases}
$$

$$
x_n = \frac{1}{\pi}\cos^{-1}(\cos(3^n\pi x_0))
$$

$$
x^* = 0,\ \frac{1}{2},\ 1
$$

periodic if $x_0 = \dfrac{q}{p}$

i.

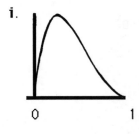

$$
x_{n+1} = 16x_n(1-2\sqrt{x_n} + x_n) \qquad 0 \le x_n \le 1
$$

$$
x_n = \sin^4(2^n\sin^{-1}x_0^{\frac{1}{4}})
$$

$$
x^* = 0,\ \frac{9}{16}
$$

periodic if $x_0 = \sin^4\left(\pi\dfrac{q}{p}\right)$

j.

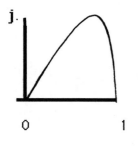

$$x_{n+1} = \sqrt{2} \, x_n(1-x_n^4)^{\frac{1}{4}} \quad 0 \leq x_n \leq 1$$

$$x_n = (\sin (2^n \sin^{-1} x_0^2))^{\frac{1}{2}}$$

$$x = 0, \quad \left(\frac{3}{4}\right)^{\frac{1}{4}}$$

periodic if $x_0 = [\sin (\pi \frac{q}{p})]^{\frac{1}{2}}$

k.

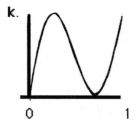

$$x_{n+1} = 16 \, x_n^3 - 24 \, x_n^2 + 9x_n \quad 0 \leq x_n \leq 1$$

$$x_n = \frac{1}{2}[1 + \sin((1-3)^n \sin^{-1}(2x_0 -1))]$$

$$x^* = 0, \quad \frac{1}{2}, \quad 1$$

periodic if $x_0 = \sin^2\left(\pi \, \frac{2q+1}{4p}\right)$

Problems

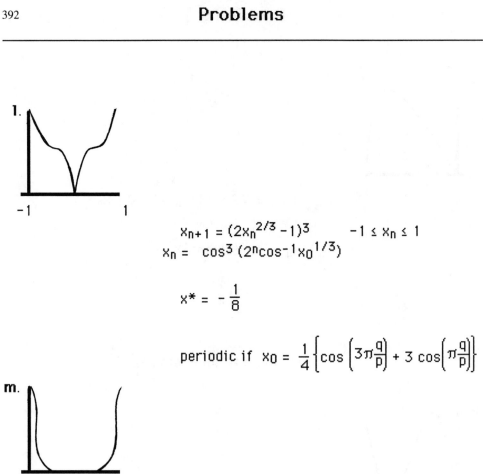

l.

$$x_{n+1} = (2x_n^{2/3} - 1)^3 \qquad -1 \le x_n \le 1$$
$$x_n = \cos^3(2^n \cos^{-1} x_0^{1/3})$$

$$x^* = -\frac{1}{8}$$

periodic if $x_0 = \frac{1}{4}\left\{\cos\left(3\pi\frac{q}{p}\right) + 3\cos\left(\pi\frac{q}{p}\right)\right\}$

m.

$$x_{n+1} = (2x_n^6 - 1)^{1/3} \qquad -1 \le x_n \le 1$$

$$x_n = (\cos(2^n \cos^{-1} x_0^3))^{1/3}$$

$$x = -\left(\frac{1}{2}\right)^{1/3}, \ 1$$

periodic if $x_0 = \left[\cos\left(\pi\frac{q}{p}\right)\right]^{\frac{1}{3}}$

(iii)   Prove in each case above that all the periodic solutions indicated become chaotic when $\dfrac{q}{p}$ is replaced by an irrational number.

(All the examples above can be considered as integrable iteration schemes because x_n can be solved by explicit formulae depending on initial conditions x_0. Nonetheless, each example illustrates chaotic dynamics as mentioned in (iii) as a small perturbation of periodic motion. An interesting problem would be: How do these explicit solutions in the above iteration schemes relate to quasiperiodic and almost periodic motion? Another interesting point relates to the perturbation of these iteration schemes. Thus in example a, when the constant 4 is perturbed to the parameter λ, the Feigenbaum map discussed in the text in Chapter 2 results. The associated iteration scheme is not integrable at all for λ in the open interval (0, 4). An interesting problem is to extend each of the iteration schemes discussed above by letting the appropriate absolute constants be denoted by a parameter λ. Question: Do the new iteration schemes with the parameter λ appropriately restricted have the same dynamics as the Feigenbaum map, namely repeated bifurcation, and period-doubling route to chaotic dynamics?)

Problems on the Initial Value Problem for Differential Equations

1)   Show the initial value problem for the system

$$y_1' = y_2 y_3 \quad , \qquad y_2' = -y_1 y_3 \quad , \qquad y_3' = -k^2 y_1 y_2$$

$$y_1(0) = 0 \quad , \qquad y_2(0) = 1 \quad , \qquad y_3(0) = 1$$

(where k denotes a constant lying between 0 and 1) is defined for all values of the argument t, and has a unique solution.

[Hint: use the contraction mapping theorem and an appropriate iteration scheme]

2) (i) Prove using 1) that $y_1^2 + y_2^2 = 1$

$$k^2 y_1^2 + y_3^2 = 1$$

(Note these equations can be regarded as conservation laws).

(ii) Deduce that the functions $y_1(t)$, $y_2(t)$, $y_3(t)$ are periodic functions.

3) The functions $y_1(t)$, $y_2(t)$, $y_3(t)$ defined in 1) are called Jacobian elliptic functions

$$y_1(t) = sn(t) \qquad y_2(t) = cn(t) \quad \text{and} \quad y_3(t) = dn(t)$$

Prove that each of these functions satisfies a nonlinear second order ordinary differential equation.

4) Prove that as the parameter $k \to 0$

$$snt \to \sin t \qquad cn(t) \to \cos t \qquad dnt \to 1$$

and that as $k^2 \to 1$ the elliptic functions degenerate into hyperbolic functions

$$snt \to tan\ ht \qquad cn(t) = dn(t) = sech\ t$$

and are no longer periodic.

5) Find the traveling-wave solution, in the form $u(s, t) = f(x-ct)$, for each of the following equations

(i) Burgers equation

$$u_t + uu_x = u_{xx},$$

with $u \to 0$ as $x \to +\infty$

and $u \to u_0 (>0)$ as $x \to -\infty$

(ii) Modified Kdv equation

$$u_t + 6u^2u_x + u_{xxx} = 0$$

with $u, u_x, u_{xx} \to 0$ as $|x| \to \infty$

6) Nonlinear Schrödinger equation. Consider the equation

$$iu_t + u_{xx} + u|u|^2 = 0$$

and seek a traveling-wave solution in the form

$$u = re^{i(\theta+nt)}$$

where $r(x-ct)$ and $\theta(x-ct)$ are real functions and c and n are real constants.

7) Fisher's equation. Seek traveling-wave solutions of the equation

$$u_t = u_{xx} + \alpha^2 u(1-u),$$

where $\alpha \to 0$ is a constant, in the form $u(x, t) = f(x-ct)$. Investigate the nature of these solutions by examining the phase-plane (f, f') and show in particular, that there exist solutions such that $f \to 0$ as $x \to -\infty$, $f \to 1$ as $x \to +\infty$, for all $c \leq -2\alpha$.

8) Burger's equation. Show that, if $u(x, t)$ is a solution of $u_t + uu_x = \lambda u_{xx}$ $-\infty < x < \infty$ where λ is a (positive) constant, then the integral of u is a conserved quantity.

9) Nonlinear Schrödinger equation. Show that, if $u(x, t)$ is a solution of the one-dimensional nonlinear Schrödinger equation,

$$iu_t + u_{xx} + \lambda u|u|^2 = 0 \qquad -\infty < x < \infty$$

where λ is a positive constant, then

$$\int_{-\infty}^{\infty} |u|^2 dx, \qquad \int_{-\infty}^{\infty} (u^* u_x - uu_x^*)dx \qquad \text{and}$$

$$\int_{-\infty}^{\infty} \left(|u_x|^2 - \frac{1}{2}\lambda|u|^4\right)dx \qquad \text{are constants of the motion. (The }*$$

denotes the complex conjugate.)

Problems on Differential Geometry

A good reference for the next few problems is the classic book by
L. P. Eisenhart, <u>Riemannian</u> <u>Geometry</u>, Princeton University Press.

10) Let a conformal deformation be defined by $\bar{g}_{ij} = e^{2\sigma}g_{ij}$. Let
$\Delta_1\sigma = |\nabla\sigma|^2$ and Δ_2 denote the Laplace-Beltrami operator. Find a
formula relating the deformed Ricci tensor $\bar{R}_{ij}$ to the original Ricci
tensor R_{ij} and the functions $\Delta_1\sigma$ and $\Delta_2\sigma$.

11) Use the formula in 10) above to verify the scalar curvature
functions R transformation properties

$$\bar{R} = \bar{g}^{ij}\bar{R}_{ij} = e^{-2\sigma}\left[R + 2(n-1)\Delta_2\sigma + (n-1)(n-2)\Delta_1\sigma\right]$$

12) A space of which $R_{ij} = \frac{R}{n}g_{ij}$ is called an Einstein space.

i) Prove 2-dimensional. Every (M_2, g) is an Einstein space. Show
that an Einstein space of 3 dimensions (M_3, g) has constant
curvature.

13) Show that a space of constant curvature K_0 is an Einstein
space, and that $R = K_0(1-n)n$.

14) If an Einstein space is conformal to a flat space, it is a space
of constant curvature.

15) Show by means of that when $n > 2$ the scalar curvature of an
Einstein space is constant.

More Problems on Iteration Schemes

1) Investigate the following iteration near <u>fixed</u> <u>points</u> and <u>given</u> <u>parameter</u> <u>values</u>. Investigate stability

    i) $x_{n+1} = (1+\mu)x_n + x_n^2$ near $x = 0$ $\mu = 0$

    ii) $x_{n+1} = -(1-\mu)x_n + x_n^3$ near $x = 0$ $\mu = 0$

    iii) $x_{n+1} = \mu + x_n - x_n^2$ near $x = 0$ $\mu = 0$

    iv) $x_{n+1} = kx_n(1-x_n)$ near $x = 0$ $k = 1$

2) What happens to the iteration process

$$x_{n+1} = kx_n(1-x_n)$$

for k near 3? How do the fixed points of the second generation map appear as k increases beyond 3.

Hilbert Space Facts

Prove the following elementary results concerning a Hilbert space H over the real numbers with inner product (,)

1) Schwartz inequality (with equality if and only if x and y are proportional)

$$|(x, y)| \leq \|x\| \, \|y\|;$$

2) Triangle inequality

$$\|x+y\| \leq \|x\| + \|y\|$$

3) Parallelogram Law

$$\|x+y\|^2 + \|x-y\|^2 = 2(\|x\|^2 + \|y\|^2)$$

Prove this law characterizes Hilbert spaces.

4) The Riesz Representation Theorem — For every bounded linear functional F on a Hilbert space H, there is a uniquely determined element $f \in H$ such that $F(x) = (x, f)$ for all $x \in H$ and $\|F\| = \|f\|$.

5) Let T denote a completely continuous linear operator defined on a Hilbert space H. Let the linear operator $L = I - T$. Prove the closure of the range of L is the orthogonal complement of the null space of L*.

Prove that the range of L is closed, so that the closure operation just mentioned is redundant.

6) A bounded sequence in a Hilbert space contains a weak convergent subsequence.

A) Let M be a closed linear subspace of H. Then for every $u \in H$ not in M there is a $v \in M$ such that

$$\|u-v\| = \text{g.l.b.} \ \|u-w\|$$

$$w \in M$$

B) . (Projection Theorem). Let M be a closed linear subspace of H. Then for every $u \in H$, $u = v+w$, where $v \in M$ and $(w, M) = 0$ [i.e., $(w, h) = 0$ for all $h \in M$.

C) (Banach-Saks). Let $\{v_n\}$ be a sequence of elements in H such that $\|v_n\| \le k$. Then one can find a subsequence $\{v_{n_j}\}$ and an element

v such that $\left\| \dfrac{v_{n_1} + \ldots + v_{n_r}}{r} - v \right\| \to 0$ as $r \to \infty$.

More Problems on Systems of Differential Equations

10)   Consider

$$(*)$$

$$\frac{dN_2}{dt} = -cN_2 + dN_1 N_2$$

$$\frac{dN_1}{dt} = aN_1 - bN_1 N_2$$

$$\Big(a, b, c, d > 0$$

(i) Investigate exactly when the solutions of (*) are periodic.
(ii) Find a conserved quantity $I(x, y)$ for (*).
(iii) Can (ii) be used to derive (*)?

11)   Prove that   sn u,   cn u   and dn u   satisfy the following differential equations

i)
$$\frac{d^2}{du^2} sn\, u = 2k^2 sn^3 u - (1+k^2)sn\, u$$

ii) Prove dn u satisfies

$$\frac{d^2}{du^2} cn\, u = (2k^2-1)cn\, u - 2k^2 cn^3 u$$

iii)
$$\frac{d^2}{du^2} dn\, u = (2-k^2)dn\, u - 2dn^3 u$$

Elementary Problems on Inequalities

1) If $y(0) = y(\pi) = 0$ and y' is L^2, then

$$\int_0^\pi y^2 dx < \int_0^\pi y'^2 dx$$

unless $y = c \sin x$

2) If y has the period 2π, y' is L^2 and

$$\int_0^{2\pi} y\, dx = 0$$

then

$$\int\limits_{0}^{2\pi} y^2\,dx \;<\; \int\limits_{0}^{2\pi} y'^2\,dx$$

unless $y = A \cos x + B \sin x$

3) If y' is L^2, the

i) $\displaystyle \int\limits_{0}^{\infty} \frac{y^4}{x^3}\,dx \;\leq\; \frac{3}{2}\left(\int\limits_{0}^{\infty} y'^2\,dx\right)^2$

unless

ii) $y = \dfrac{x}{ax+b}$

where a and b are positive, in which case there is equality.

4) Show the equation

$$\nabla^2 h - \frac{h}{\lambda^2} = -\frac{\Phi_0}{\lambda^2}\,\hat{z}\,\delta_2\,(r)$$

has the exact solution

$$h(r) = \frac{\Phi_0}{2\pi\lambda^2}\,k_0\!\left(\frac{r}{\lambda}\right)$$

where k_0 is a zero-order Hankel function of imaginary argument.

5) Prove $k_0\left(\frac{r}{\lambda}\right)$ cuts off as $e^{-r/\lambda}$ at large distances and diverges logarithmically as $\ln\left(\frac{\lambda}{r}\right)$ as $r \to 0$.

Problems on Inequalities Connected with Simple Elliptic Partial Differential Equations

1) For any smooth function $u(x_1, x_2, \ )$ of compact support in R^3, the inequality

$$\int\limits_{-\infty}^{\infty}\int u^4 dx_1 dx_2 \leq 2\int\limits_{-\infty}^{\infty}\int u^2 dx_1 dx_2 \int\limits_{-\infty}^{\infty}\int grad^2 u \, dx_1 dx_2$$

holds.

2) For any smooth function $u(x_1, x_2, x_3)$ of compact support in R^2, the inequality

$$\iiint\limits_{-\infty}^{\infty} u^4 dx_1 dx_2 dx_3 \leq 4\left(\iiint\limits_{-\infty}^{\infty} u^2 dx_1 dx_2 dx_3\right)^{\frac{1}{2}}$$

$$\left(\iiint\limits_{-\infty}^{\infty} grad^2 u \, dx_1 dx_2 dx_3\right)^{\frac{1}{2}}$$

holds.

3) Prove Young's inequality

$$ab \leq \frac{a^P}{P} + \frac{bP'}{P'} \qquad \left(\frac{1}{P} + \frac{1}{P'} = 1 \ ; \quad P.P' > 1 \right)$$

the inequality. Prove

$$\int\limits_{-\infty}^{\infty}\!\!\int u^4 dx_1 dx_2 \leq \Sigma \left(\int\limits_{-\infty}^{\infty}\!\!\int \text{grad}^2 u \, dx_1 dx_2 \right) + \frac{1}{\Sigma} \left(\int\limits_{-\infty}^{\infty}\!\!\int u^2 dx_1 dx_2 \right)^2$$

which is valid for any $\Sigma > 0$, and (3) implies

$$\int\limits_{-\infty}^{\infty}\!\!\int\!\!\int u^4 dx_1 dx_2 dx_3 \leq \frac{1}{2^3} \left(\int\limits_{-\infty}^{\infty}\!\!\int\!\!\int u^2 dx_1 dx_2 dx_3 \right)^2 + 3\Sigma \left(\int\limits_{-\infty}^{\infty}\!\!\int\!\!\int \text{grad}^2 u \, dx_1 dx_2 dx_3 \right)^2$$

for any $\Sigma > 0$.

1) For any smooth function $u(x_1, x_2, x_3)$ of compact support, the inequality

$$J(u) = \int\limits_{-\infty}^{\infty}\!\!\int\!\!\int u^b dx_1 dx_2 dx_3 \leq 48 \left(\int\limits_{-\infty}^{\infty}\!\!\int\!\!\int \text{grad}^2 u \, dx_1 dx_2 dx_3 \right)^3$$

holds.

2) For any function $u(x) \in \overset{\circ}{W}^2_1(\Omega)$, we have

$$\int\limits_\Omega u^2 dx \leq \frac{1}{\mu_1} \int\limits_\Omega grad^2 u dx$$

Problems on Completely Continuous Linear Operators in a Hilbert Space

Assume that T is a completely continuous linear operator defined on a Hilbert space H. This means that whenever $\{u_i\}$ is a bounded sequence in H, the sequence $\{Tu_i\}$ has a convergent subsequence. Let $L = I - T$ and set ker L to be the set of all $u \in H$ which satisfy $Lu = 0$

1) Prove ker L is a finite dimensional subspace of H.

2) Prove there is a constant $c_0 > 0$ such that $\|Tu\| < c_0 \|u\|$ for all $u \in H$.

3) Prove the equation $Lu = f$ has a solution if and only if $f \in M^*$, where $M^* \perp$ ker L^*. Here L^* denotes the adjoint of the operator L.

3a) Prove $L^* = I - T^*$ where T^* is the adjoint of the completely continuous operator T. In addition, prove T^* is completely continuous.

3b) Prove the operator $L = I - T$ is a Fredholm operator of index zero. Prove ker L and ker L^* are of the same finite dimension.

4) Consider the iterate L^n for an arbitrary positive integer n. Prove L^n is of the form $I - T_n$ where T_n is a completely continuous operator. Prove the Fredholm index of the operator L^n is zero.

More Problems Concerning the Local Geometry of a Two-Dimensional Surface S

0) Let the first fundamental form of a surface S be denoted

$$ds^2 = E\, du^2 + G\, dv^2$$

Verify that the Gauss curvature of the surface S can be written

$$K = -\frac{1}{2\sqrt{EG}}\left[\frac{\partial}{\partial u}\frac{G_u}{\sqrt{EG}} + \frac{\partial}{\partial v}\frac{E_v}{\sqrt{EG}}\right]$$

1) If $P(u, v)$ and $Q(u, v)$ are two functions of u and v on a surface, then, according to Green's theorem and the expression for the area of a small rectangle $dA = (EG - F^2)^{1/2}\, du dv$

$$\int_c P\, du + Q\, dv = \iint_A \left(\frac{\partial Q}{\partial u} - \frac{\partial P}{\partial v}\right)\frac{1}{\sqrt{EG-F^2}}\, dA$$

where dA is the element of area of the region R enclosed by the curve C on the surface S. Show the geodesic curvature of the curve C satisfies the following formula, where at a point the curve C makes an angle θ with the coordinate curve v = const.

$$\int_c k_g \, ds = \int_c d\theta + \iint_A \left(\frac{\partial}{\partial u}\left(k_2\sqrt{G} \right) - \frac{\partial}{\partial v}\left(k_1\sqrt{E} \right) \right) du \, dv$$

Here k_1 and k_2 are the geodesic curvatures of the curves $v =$ constant and $u =$ constant respectively. Moreover,

$$\int_c k_g \, ds = \int_c d\theta - \iint_A K \, dA$$

Thus we find that if C is a smooth curve C which bounds the region R on which there are no points where the slope of C has discontinuities, we find the simple Gauss–Bonnet formula

$$\int_c k_g \, ds + \iint_A K \, dA = 2\pi$$

Indeed, in this case,

$$\int_c d\theta = 2\pi$$

More Problems Concerning Nonlinear Differential Equations

Problem   Consider solutions of the equation

(*) $\ddot{y} + \lambda \exp(y) = 0$

which also satisfies the two-point boundary condition: $y(a) = y(b) = 0$. Assume that $a = 0$ and that $\lambda > 0$.

Prove the following

Fact For every value of λ taken between 0 and λ_1, where

$$\lambda_1 = \frac{(1.8745...)^2}{b^2}$$

equation (*) has two real and distinct solutions whose graphs in the (x, y) axis are called C_1 and C_2. These curves C_1 and C_2 are of parabolic form, concave to the x-axis and pass through the points $x = 0$ and $x = b$. Can you find explicit representations for the curves C_1 and C_2?

As $\lambda \to \lambda_1$, prove the two curves tend toward a limiting curve C. Prove when $\lambda = \lambda_1$, the limit curve is the unique solution of equation (*). For $\lambda > \lambda_1$, prove the boundary value problem has no real solution.

Does the equation (*) exhibit the nonlinear desingularization phenomenon as $\lambda \to 0$?

Problem Consider the smooth solutions of the two-point boundary value problem

(1) $$\frac{d^2u}{dt^2} + \lambda(u - u^3) = 0 \; ; \quad u(0) = u(\pi) = 0$$

relative to the solutions of linear eigenvalue problem

$$\frac{d^2u}{dt^2} + \lambda u = 0 \; ; \quad u(0) = u(\pi) = 0$$

Denote the eigenvalues of this last boundary value problem by λ_i where one has the ordering $0 < \lambda_1 < \lambda_2 < \lambda_3 < \ldots$

Show

(i) If $0 \leq \lambda \leq \lambda_1$, $u(t) \equiv 0$ is the unique solution of (1).

(ii) Prove $(0, \lambda_1)$ is a bifurcation point for the boundary value problem (1).

(iii) Prove $(0, \lambda_i)$ is a bifurcation point for the boundary value problem (1).

(iv) Use the elliptic functions introduced above to make these results (i – iii) concrete.

(v) Does the boundary value problem (1) have any other bifurcation points?

(vi) Relate this problem to the two-dimensional problem

$$\Delta u + \lambda(u - u^3) = 0 \quad \text{defined inside the rectangle R with}$$
$$\text{sides a and b}$$

where u vanishes on the boundary of the rectangle R.

(vii) What happens when $a = b$ in (vi)?

Problems Concerning Periodic Solutions of Ordinary Differential Equations

The simplest example of an ordinary differential equation not integrable by quadrature, in the traditional sense, is the Riccati equation

$$\text{(8)} \qquad \frac{dy}{dt} + y^2 = f(t)$$

To study this equation by our methods we focus on the left-hand operator

$$\text{(9)} \qquad A(y) = \frac{dy}{dt} + y^2$$

and instead of considering the initial value problem we focus on T-periodic boundary conditions

$$\text{(10)} \qquad y(0) = y(T)$$

As discussed above, in Chapter 1, the singular points of the operator A turn out to be a hyperplane of co-dimension one in the Sobolev space $W_{1,2}(0, T)$. The singular values of the operator A turn out to be the boundary of an infinite dimensional convex set in the Sobolev space. The associated operator A is a proper map. In fact all the steps outlined in Chapter 1, in our systematic procedure for integrability, can be carried out explicitly. This is described in the papers of the bibliography for Chapter 1.

The final result is a global normal form for the operator A, and explicit coordinate transformations to achieve this normal form. The result is as follows:

Theorem   The mapping A defined by equation (9) together with the periodic boundary conditions (10) is C' equivalent to a global Whitney fold by explicit coordinate changes.

RESEARCH PROBLEM

After the famous Riccati equation, the next simplest example of a system not integrable by quadrature is the Abel equation which can be written

$$(11) \qquad\qquad \frac{dy}{dt} + P_3(y, \lambda) = f(t)$$

To set up this problem in our context, assume $f(t)$ is T-periodic, and limit attention to T-periodic solutions, define an operator A as follows:

$$(12) \qquad\qquad A_\lambda y = \frac{dy}{dt} + \lambda y - y^3 , \; y(0) = y(T)$$

This operator once again has a large bifurcation set, and in fact this set (singular points) consists of points all of which are folds and cusps. What is interesting here is that this operator is integrable in our new extended sense. In fact, one can prove the following, under appropriate restrictions

Theorem A_λ, (defined by equation (12)), is C^1 equivalent to a global Whitney cusp, where the Hilbert spaces X and Y are the same as above.

Problems

ANOTHER RESEARCH PROBLEM

Study the equation

(13)
$$\frac{dy}{dt} + P_N(y, \lambda) = f(t)$$

where P_N is a polynomial of degree N in y, that depends smoothly on the real parameter λ, where the forcing function f is smooth and T-periodic. Here one studies only T-periodic smooth solutions. Together with the T-periodic conditions, define the operator $A_\lambda y$ to be

(14)
$$A_\lambda y = \frac{dy}{dt} + P_N(y, \lambda)$$

as a mapping between the Sobolev space $W_{1,2}(0, T)$ into $L_2(0, T)$, together with the periodic boundary conditions. Prove this mapping has the linearization

(15)
$$A_\lambda'(y)v = \frac{dv}{dt} + P_N'(y)v$$

$$v(0) = v(T)$$

Prove this operator is a Fredholm operator of index zero. Prove this mapping A_λ has singular points exactly when

(16)
$$\int_0^T P_N'(y(t))dt = 0$$

Research Problem: In case the degree of the polynomial P exceeds 2, find the singular values for the associated mapping A. Determine cases in which A has a global normal form. Is the normal form stable? As the degree N increases, classify all local singular points beyond Whitney folds and cusps. (Note in all cases $A'(x)y$ has a kernel of dimension at most one.)

Index

Abrikosov ansatz, 355
almost periodic, uniformly, 115, 137
almost periodic functions, Besicovitch, 137, 138
Arnold, V. I., 5, 30, 80, 84

Banach space, 21-26, 31, 72, 78, 119, 203, 236
BCS theory, 357
Bessel functions, 30, 214, 312, 333-4, 367, 370
best constant for Sobolev imbedding, 271
bifurcation diagram, 148, 178
bifurcation equation, 224
bifurcation point, 24, 29, 69, 71, 123, 203, 224, 226, 336
bubble type vortex breakdown, 328
Burger's equation, 38

Calabi conjecture, 285
calculus of variations 149
canonical mapping, 166
canonical coordinates, 12
canonical transformation 3
capacity of a set, 345
chaotic behavior, 1, 66, 68, 78, 82, 318
chaotic dynamics, 104, 114, 144
chaotic mapping, 166
Chern class, 276, 281
closed geodesic, 195, 198, 228-230
codimension, 33, 35, 72, 108, 123, 221
coerciveness, 248-9, 292, 378-9
combination tones, 80
compact manifold, 242-6, 251, 272, 274
complete integrability, 3, 8, 19-21, 28-30, 41, 43, 62, 71
completely continuous, 135, 153-8, 205, 211
complex manifold, 275
conformal deformation, 242
conservation law 7
conserved quantity 4
contraction mapping, 162, 209

convex function214
coupled dynamic systems5
covariant derivative, 349
critical points 149
cusp, 71-73

diagonal map 5
diagonalization 5
diffeomorphic4
differential inequality, 201
Dirac delta function, 160, 212
Dirichlet integral, 160

Einstein metric, 274-8
Euler-Lagrange equations, 97, 152, 183, 201, 228, 276, 286, 307,
       331
exchange of stability, 164-5

Feigenbaum map, 164-6
flat convergence, 239
flow force, 333-9
flux quantization, 353-4
fold point, 60, 72
fold singularities, 67, 69
folds, infinite dimensional, 147, 175, 221
Fourier series, 33, 36, 79, 82-4, 88, 113-17, 410, 412
Frechet derivative, 8, 25, 118, 153, 173-6, 216
Fredholm alternative, 119, 172, 226, 246-7, 285, 411
Fredholm mapping property, 25
Fredholm operator, 21, 25, 31, 119-20, 147, 174, 216, 221
Fredholm operator, nonlinear, 25, 32, 50, 69-70, 118, 120, 123-4,
       145, 147, 216

gauge invariant, 349, 352, 359, 361
gauge transformation, 349
Gauss-Bonnet theorem, 199-200, 232-3, 247, 252, 274, 282
Gaussian curvature, 242, 247, 258, 266, 275, 277
generalized solution, 131-5, 248-9, 272, 313-14, 380
geodesic curvature, 201, 233-4
Ginzberg-Landau action functional, 349, 351, 363
Ginzberg-Landau equations, 349-51, 360-3, 371, 376-7, 381
Ginzberg-Landau functional, 353
 global fold, 36-7
global homeomorphic, 173
global linearization, 8, 19-20, 28, 38, 427
global normal form, 22, 27, 40, 71-2, 147, 175, 213, 220, 409-13

gradient map 151
gradient operator, 149, 151-4, 158, 229, 286, 289, 313-16
Green's function, 128, 179-85, 259-60, 302, 342-7, 350, 355, 367,
       369

Hadamard's theorem, 45
Hamiltonian-Jacobi, 419
Hopf, H., 122, 257
helicity, 301, 310-11
Helmholtz equation, 350, 354, 370
Helmholtz circular vortex filament, 309,342
high Tc superconductivity, 356-7
Hill's equation, 35
Hills's spherical vortex, 308
Hodge theory, 228

ideal fluid, 249
incompressible fluid, 297
index, Fredholm, 119
infinite-dimensional manifold, 37
integrability, complete, 3,19
integral current, 237
integration by quadrature, 12

inverse scattering method, 9
involution, 3
irrotational vector field, 294
iteration, 160
I-V curve, 224

Jacobi, 1
Jensen's inequality, 215
Jordan canonical form, 14

Kahler deformation, 275
Kähler manifold, 278
KAM theory, 39
Kondrachov compactness criterion, 133
Korteweg de Vries equation, 7

Lagrange multiplier, 95
Laplace operator, 127,130
Lax-Milgram theorem, 45
leapfrogging of vortices, 317-25
Leonardo da Vinci, 294
Liapunov's theorem, 85-92
Liapunov-Schmidt reduction, 207
linear Dirichlet problem, 127
linearization, global, 8
Liouville's theorem, 4
Ljusternik-Schnirelmann theory, 89
London equation, 359

material rate of change, 300
Maxwell equations, 348
mean curvature, 257-60
Meissner effect, 356
minimax determination, 21
Monge-Ampere operator, 280

Monge's problem, 253
Morse inequalities. 167, 193

natural constraints, 194-7
nonlinear analysis, 213
nonlinear desingularization, 176-90
nonlinear differential operator, 19-62
nonlinear Dirichlet problem, 40-62, 178-184
nonlinear dynamics, 66, 101, 104, 330
nonlinear eigenvalue problems, 144,147,153-7,177, 185, 257- 73,
      307-14, 322-9, 351
nonlinear parabolic partial differential equations, 28, 67, 222
normal current, 236
normal mode, 76, 89, 92

period doubling bifurcation, 165
perturbation, 1-40
points, regular, 70, 124, 174
point, singular, 21-36, 43-74, 120-3, 147, 173, 213-228
Poisson bracket, 3
Poisson equation, 241

quadrature, integrability by, 1,17-37
quasiperiodic, 2-40, 74-117

rank normal form, 13-17, 41
Rellich's lemma, 133
Riccati equation, 2, 17, 28, 71, 174
Riccati operator, 17, 28, 71, 174
Ricci tensor, 274-293
Riemannian metric, 228-293
Riesz representation theorem, 134-136

scalar curvature, 261-68

second kind of Bessel function of zero order, 370
self-adjoint operator 17-118
self-duality, 350,353-5
semiconductor device, 222-227
sine-Gordon equation 257-260
singular values 21-37,47
small denominators 75
Sobolev space 32 -36, 127-136 ,242-293,304-316,324-6,371-380
Sobolev imbedding theorem 128-136,271
Sobolev inequality 133,271
solenoidal vector field 294,312
soliton10,299,302,317
solution diamond, 136
spectral theory,35,228
spiral type vortex breakdown, 326,337
Squire parameter, 329
Squire-Long equation, 326-341
steepest descent method, 167-17,266-73
Stokes stream function, 297-347
subcritical flows, 326-341
superconducting vortices,348-380
supercritical flows, 326-341
swirl, 304-341
switching behavior, 222-227
symmetrization, 182,315.

Thyristor, 326-41
Toda lattice 6,7
topological singular set 62-65
topological singular set image 62-65
torus,4,74,102,111,228
trigonometric polynomial, truncation 137,138
turbulence 330
type II superconductivity, 350,353,356-358

uniformization theorem, 252-4
universal covering surface, 242

vector potential, 301
viscosity, 295,297
vortex atoms,295
vortex breakdown, 326-341
vortex filament, 300,302,309-10,316-7,342-7
vortex number, 352-5,359-367
vortex ring, 304-347
vortex strength, 300
vorticity of a fluid, 294
vorticity equation, 299,362

weak convergence, 98,150-153,292,378
Whitney cusp 73,126
Whitney fold,26-7,33-7,62,71-2,122,147,213-226

Yamabe problem, 261-273
Yang-Mills theory,348-352